Flat Out
And
Half Turned Over

Flat Out
And
Half Turned Over

TALES FROM PIT ROAD
with Buddy Baker

By Buddy Baker

with David Poole

SPORTS
PUBLISHING

Sports Publishing books may be purchased in bulk at special discounts for sales promotion, corporate gifts, fund-raising, or educational purposes. Special editions can also be created to specifications. For details, contact the Special Sales Department, Sports Publishing, 307 West 36th Street, 11th Floor, New York, NY 10018 or sportspubbooks@skyhorsepublishing.com.

Sports Publishing® is a registered trademark of Skyhorse Publishing, Inc.®, a Delaware corporation.

Visit our website at www.sportspubbooks.com.

10 9 8 7 6 5 4 3 2 1

Library of Congress Cataloging-in-Publication Data is available on file.

ISBN: 978-1-61321-355-1

Printed in the United States of America

For Buck Baker and all of the racing heroes just like him who helped drive NASCAR through its early days. They will never get enough credit for laying the foundation the sport keeps building on today.

—David Poole

Contents

Introduction

Who's Your Buddy?

When it comes to telling great stories, Buddy Baker has just about everybody else a lap down.

When I took over as motorsports writer for *The Charlotte Observer* in 1997, I quickly learned that one of the best ways to spend a race-day morning was to go to where the laughter was. Whenever Buddy was around for a CBS or TNN telecast, he could be found in the press box hours before the green flag, telling stories.

Every time he came to the punch line, he'd let out a laugh that you could recognize from a county away. Before long, I'd heard some of the same stories several times. Somehow, though, they were just as funny every time I heard them.

The original idea for this book was to go around the NASCAR garage and collect the best stories from different drivers. That's still a fine idea, but there was something about Buddy's stories that made me want to focus on him this time.

Maybe it's because we all miss having Buddy around as much as he used to be before NASCAR auctioned off its television rights to Fox Sports and NBC and their cable partners. Buddy didn't get a spot on those broadcast teams, and while people like Darrell Waltrip, Larry McReynolds, Benny Parsons and Wally Dallenbach are doing a fine job, a lot of fans still wish the networks could have found a place for Buddy.

He's still doing television, working American Speed Association races for TNN, and still makes it to some NASCAR races for speaking engagements for a tour company that brings groups to major events. I hope this book won't cut into Buddy's material for those speeches.

It's hard to have a lot of good stories if you haven't had much of a life. For Buddy, that's not a problem. He grew up around racing because of his father, two-time NASCAR champion Buck Baker, who won 46 races in what was then called the Grand National division before retiring in 1976 after 636 career races.

Buck Baker was an original, a gruff old cuss who asked for no quarter on the track and minced no words off it. One of my favorite stories involving the Bakers happened in Daytona Beach, Fla., in February 1998 when NASCAR was celebrating its 50th anniversary by naming the 50 greatest stock-car drivers of all time.

Buck and Buddy were both on that list, as they should have been. During a ceremony held at the Daytona USA attraction, each surviving member of this elite group of drivers was introduced and presented with a new leather jacket signifying his selection. As each driver took a seat at the front of the auditorium, he put on his jacket and sat waiting for the rest of the group to be introduced. Dozens of photographers were lined up to take group shots at the end of the ceremony.

Since the drivers were introduced alphabetically, the Bakers were among the first to be brought out, and they sat side by side in seats on the front row of the dais. As the introductions continued, I noticed that Buck had his leather jacket folded neatly and sitting in his lap. Everyone else, including Buddy, had dutifully slipped his on.

Buddy elbowed Buck in the side, trying to cajole his father into getting into uniform. Buck shooed Buddy's advice away like he was swatting a mosquito. Finally Buddy gave up. As soon as the group photos were done, I walked up to Buddy and he just started laughing.

"What'd Buck say about the jacket?" I asked.

"He said, 'The jacket I've got on looks nicer than that one; I'm not putting that thing on to have my picture taken,'" Buddy said. "I tried to tell him that was the whole idea, but I know when I'm wasting my time."

When Buck died earlier this year, a writer for another newspaper was talking to Buddy for the obituary. Understanding it was a difficult time for the family, the reporter was trying to be as sensitive as he could be in the conversation.

"Is it fair to say your father was a little…crusty?" the reporter said.

Buddy laughed.

"A little?" he said.

"He was Buck!"

Buck was Buck and Buddy is Buddy.

As you will read in the pages that follow, Buddy loves racing as much as he loves life. That's where the title comes from. During one of our conversations, Buddy said, "Everything I did in my life I was flat out and half turned over."

It's too bad that a lot of the new fans NASCAR has collected in the past decade recognize Buddy Baker more as a television analyst than as a racing champion. Buddy strongly believes that many of racing's great heroes of the past don't get the kind of recognition they deserve today, and that's at least partially true about Buddy himself.

Buddy's first race in NASCAR's top series was at Columbia Speedway in 1959. He was 18. It took him more than 200 races before he finally got his first victory, but during those early years Buddy says he learned a little bit of something from every car owner he drove for and every competitor he drove against.

When he finally got into top-quality cars, Buddy emerged as a star—particularly on the sport's biggest tracks. He was the first driver to win at each of the sport's "Big Four" tracks— Daytona, Talladega, Darlington and Charlotte. He won four races at Talladega, four at Charlotte, and two each at Daytona and Darlington—meaning he got 12 of his 19 career victories on the sport's biggest stages.

He won 40 poles. He finished second 43 times and third 58 times. He had 202 top-five finishes and 311 top-ten finishes. From 1967, the year of his first win, through 1983, the year he won for the last time, Buddy had 210 top-ten finishes in 383 races.

Buddy raced more than 181,000 miles in a Winston Cup car —and probably traveled twice that many miles in his life to do it.

He ran 151,129 laps and led 9,748 of them, leading at least one lap in 242 different races.

For all that hard work, Buddy earned a career total of $3,995,500. Ask him how much he could have earned in today's era with a record like that and Buddy just laughs that great big laugh of his.

When he finally won the Daytona 500 in 1980, after coming close many times, Buddy averaged 177.602 mph for 500 miles to set a record that still stands in the sport's biggest race. He was also the first driver to ever run an official lap at more than 200 mph, accomplishing that at Talladega in March of 1970. No wonder one of the many nicknames he had was "Leadfoot."

Buddy was inducted into the Court of Legends at Lowe's Motor Speedway in Charlotte in 1995 and two years later was inducted into both the National Motorsports Press Association's Stock Car Racing Hall of Fame in Darlington, S.C. and the International Motorsports Hall of Fame in Talladega, Ala.

He later earned a place in the living rooms of race fans all over the country with his work on television, helping race fans understand what it was like to be in position to win the Daytona 500 or battle three-wide at 200 mph at Talladega.

Even at 61 years old, he's still climbing into race cars, regularly helping Rusty Wallace and up-and-coming star Ryan Newman test their Fords. Buddy's not sure he could still win a race, but he's convinced there are days he could still run in the top ten. I don't doubt it for a minute.

Last year, Buddy went to Talladega to help Wallace and Newman out with a test. At one point in the session, Buddy had turned the fastest lap of the day in one of Wallace's cars, running a lap at nearly 190 mph with a restrictor plate motor under the hood.

One of the sport's current crop of young drivers came up to Buddy and decided to see if he could do a little needling.

"Let me ask you something," the driver said. "Do you really think you could go out there and run with us?"

Buddy studied the question for a minute.

"Son, I really don't know," Buddy finally said. "I've never run this slow here."

—David Poole, July 2002

Editor's Note: Poole's original introduction, like the rest of the text, has been preserved. Though it is important to note that Baker is now 72 and can be heard on Sirius XM's *The Late Shift.*

1

Buck

Family is perhaps NASCAR's greatest tradition. The Pettys, the Allisons, the Jarretts and the Earnhardts have left their marks on stock-car racing, and the Bakers have been right there with them every step of the way.

Buck Baker drove in the first race held in what is now NASCAR's Winston Cup series on a three-quarter-mile dirt track in Charlotte in 1949. He won the 1956 and 1957 championships in what was then called the Grand National division, winning 24 races in those two seasons. In 636 career starts, Buck won 46 races and 44 poles.

Buck was a no-nonsense kind of driver and a no-nonsense kind of guy. He was also his son's hero.

My father was a great athlete. He'd find a bicycle and get on it backward and ride it around in the yard. He'd pick up the back end and ride on the front tire. He was a tremendous baseball pitcher who could bat and throw either way. You should have seen Dad shoot pool. He was one of the best I've ever seen.

I was born in Florence, S.C., and we lived in Rock Hill, S.C., for a couple of months after Dad got out of the navy. Then we moved to Charlotte. He became a bus driver, which later led to people in racing writing about him being the "flying bus driver."

He had a load of passengers on his bus going to Columbia, S.C., one time, and he wanted to go to this dance down near Great Falls. So he took a head count. Who wanted to go to the dance and who didn't? He took the whole busload to the dance and then on to Columbia. They were looking all over the place for this bus.

Buck in his winning car, a 1952 Hudson Hornet, after a race at Columbia Speedway in South Carolina on April 12, 1952. Dad won $1,000 in the 200-lap race.

When he drove the city bus around Charlotte it was always loaded down. He'd win the race at the old Charlotte fairgrounds, and the people would get on the bus the next week hoping to talk to him about racing.

When Dad was promoting the races at the fairgrounds, one time he won 27 races in a row. And people were booing him. I didn't understand that. To me, he was the greatest guy in the world.

We'd get to the track, though, and the fans all wanted to see somebody else besides the promoter win a race. They would start booing him.

I would throw stuff at people and run like the devil. I just couldn't stand it, it would gripe to me death.

Then I started looking at the people who were booing, and I realized they were the same people who'd show up and hang around the race shop on Monday or Tuesday. It was just the thing to do. They were booing just to get a reaction.

I loved all kinds of sports as a kid. We played baseball and football all around the neighborhood.

I had this bicycle, and being Buck Baker's son, I was supposed to win all of the bike races. Back then I wasn't the biggest kid in the neighborhood. In fact, I was one of the smallest, so it wasn't easy. We'd race our bikes on this little track we had built, and every kid within 20 blocks would show up. We'd have heat races and everything.

One time I knocked the pedal off on the left side of my bike, and I didn't have the money to run right out and buy a new pedal. Without that pedal on the left side of the bike, I discovered I could lay it almost all the way down on the ground. I'd take my foot and put it on the back side of the inside pedal. When the other kids would lay their bikes down, they could only go so far. I could lay mine down all the way and go right by them. I won races all summer long that year. When I finally got my new pedal, I never won again.

When I was just a little fellow I started going with Dad to the race tracks. I would go up to the other drivers and say, "My daddy's going to beat you today."

They'd say, "Get out of here!" and kick at me to run me out of their pits.

When Dad raced on the beach course at Daytona, I would go down there and watch with a group of buddies. When he'd go down the road side of the course, we'd go over there and watch him. We'd all run over to other side then and watch him come down the beach. We did as many laps like that as he did.

The cars would run down the beach, then cross it up to make it through the turn back to the road. Drivers used to talk about finding a guy with a red shirt standing on the bank and using him to mark where they should start to turn and slide into the corner. The guy in the shirt would start walking around to get a better position to see the cars, and the next time through the drivers wouldn't know he had moved. They'd turn too late and go sliding right on out of there.

The windshields would get this goo on them. You'd have ocean spray on your car. You'd turn the windshield wiper on it would just smear all over the place. Back then, the windshield wipers were 4 inches off the windshield anyway. By halfway through the race, nobody could see anything. The drivers would find spots or marks where

That's me in my father, Buck's, arms at our house in Florence, S.C., before we moved to Charlotte.

they'd start sliding to slow down, because the braking systems back then lasted about a half a lap and the rest of the time you slowed down by sliding the car.

Dad put vinegar in a foot pump, and he'd pump that up onto the windshield and cut that film. He'd be the only person who could see, but then his engine would blow up or something.

Dad once ran 140 mph in an Indianapolis car on the beach there. I still remember him telling me that if he'd hit a teacup full of sand out of place the car would still be flipping. Cars just didn't run that fast back then, especially on sand on about a four-inch tire.

Dad won the championship when NASCAR had an Indy car division, which a lot of fans don't know NASCAR ever even had. He could drive anything— modifieds, midgets, sprint cars.

When he first started racing, he had this guy named Brad working on the car, and he was a great mechanic. If we didn't win, he would sit in the back of the truck the whole way home and not say a word. If we won, he had to get drunk. And Dad was winning pretty regular, so he was getting worse and worse.

Chick Morris was driving the car towing the race car one night, and Brad was in the back seat, having already had about 15 beers. He just fell asleep in the car. If Chick was driving and it got dark, he was bad about going to sleep at the wheel. This particular time he nodded off, went off a bank, and turned end over end, pulling the race car. He wound up way out in the woods somewhere.

Here we came along in the other car following them. We saw where they went and got out there, and Chick was changing a right-front tire, saying that everything was going to be all right.

Dad asked him what happened.

"Buck," he said, "I know you're going to fire me anyway but I have to tell you something funny.

"When we turned over and stopped up here in the woods, after turning over and the seat was completely up on top of Brad, Brad raised up and said, 'Can we get a beer here?'"

Dad was going to Florida one time, and I was in the back seat. I was just a little thing. We pulled up to a gas pump on the right side of the car, looked up, and saw our race car rolling past us on the left side. The tow bar had broken in two, and the car just rolled up and stopped in the driveway there.

How lucky was that?

2

Mother Margaret

———————

Buddy's mother, Margaret, was the first of Buck Baker's wives. While Buddy eventually followed in his father's footsteps as a racer, there was also a part of Margaret in him, too.

My mother and father were married during World War II time, and I've heard her talk about hoping to have enough gas rationing stamps to go back home. She worked as a bookkeeper for the FBI in Washington, and Dad was in the navy and stationed in Maryland, so they could see each other. She played basketball while she was working up there.

Her family was in Florence, S.C., and she told me about having to get four or five patches put on the inner tube of her tire and hoping to get enough gas to get back to Washington after coming home to see them.

Dad was tough, but she was no pushover.

*My mom, Margaret, and me with the sprint car that dad
drove when NASCAR had a division for those kinds of racers.*

Dad bought this Oldsmobile that he was going to turn into a race car. It had a big motor, a lightweight body, and all of that stuff right from the factory. We were going to see my grandmother. This was back when there was almost no traffic on certain roads.

Dad made the crucial mistake of jumping my mom in his new car, wanting to blow on past her down the road. She was driving a Cadillac, and it was as fast as anything there was on the road.

Mom was a tremendous driver. She would pull the race car when Dad got sleepy. She'd tow the car all night long, get to the track, help him unload the car and then go score him during the race. Then she'd get back in the car and drive him home because he'd won the race. They were quite the team.

Dad tried to start this race with her on the way to Florence that day, and she just went up there and blew right back by him. But she didn't want to make him mad so she started to back off.

Me, being the little competitive fool that I was, I jumped in the floorboard and held her foot on the accelerator. We went by the yard where my grandmother and all the uncles and aunts were sitting around, and she and I were 15 or 20 car lengths ahead of Dad. I reached up and blew the horn to make sure everybody saw us.

Dad was mad. We were there all weekend and he never spoke to a soul.

I was about 10 when they divorced. That's tough for everybody. You have to choose which one you're going to stay with.

I lived with Mom to go to school, then I would stay with Dad for a couple of months in the summer. I would go to the races with Dad. He traveled so much that it was tough anyway. You'd have father-son outings and he would be gone.

Racing can be tough on people, but maybe that helped me get a little stronger and understand that things couldn't always go my way.

3

Let's Go Racing

Buddy spent a couple of years attending military academy in Camden, S.C. before returning to Charlotte. He attended Central High and then, in the first year the school was opened, graduated from Garinger High. He played sports and chased girls and did what high school kids did in the late 1950s.

But not every kid had a racing legend like Buck Baker for a father. And while Buddy had spent a lot of time running up and down the roads and hanging around race tracks, he really hadn't thought much about being a racer until the summer after he finished high school.

Then the genes kicked in.

I worked for my dad all that summer after I turned 18. I didn't let him know it, but I went out and ran in a couple of modified races. They were those old flat-head Fords, and I ran them a couple of times at Columbia Speedway.

Dad was so stern, and even though I worked with him, I didn't want to have the conversation about me getting started in racing. I had a feeling he would just say, "Absolutely not."

I knew that if I kept doing it he'd find out about it, though. So one day I went up to him and said, "Dad?"

He snapped back, "What?"

I said to myself, "No, I'd better not bring this up, not right now."

I went back and worked a while and walked back up to him.

I said, "Dad?"

Dad gave me my start in racing, but once I had my feet wet he told me it was time to see if I could make it on my own. Later on in life, I was glad he did it that way.

He said, "What is it?!?"

I thought, "Dang, now he's worse than he was before."

Finally, I walked up to him again and he just stopped me.

He said, "What is it!? What do you want?!"

I said, "Dad, I would like to start racing."

He simply said, "See that old car over there in the corner? The parts are all there; just get some people and y'all put it together, and you can run it."

I got a bunch of guys pretty quick. Back then, all you had to do was whistle and you had 100 people who were willing to work on your car. They were pretty good, and things weren't specialized the way they are now. Everybody could do everything.

We got her together and borrowed a car to pull her to Columbia. The race car was an old convertible that had a top you could bolt on so you could run it either way, in the Grand National division or the convertible series that NASCAR had.

That was my first race in what they called Grand National, at Columbia in 1959. I can remember starting out in that race, running dead last. I was all over the race track, and I was blaming that car.

"This is the biggest piece of junk I've ever sat down in in my entire life," I said.

Dad's car blew up, and he motioned for me to come in. I got out, and he got in that same car and made up two or three of the laps I was behind.

I said, "I've got some stuff to learn."

Dad let me run several races for him that year. Then he just said, "You know, I've introduced you to the racing world. Now it's time for you to go out and earn your own way."

We'd run at Greenville, Savannah, Columbia, and other tracks where they had races every week. It was tough to make a living driving, because nobody was paying anything, but nobody else was making any money, either. There just wasn't a lot of money around.

Somebody would say, "We can't pay you but 20 percent." I would say, "That's 20 percent more than what I've got now."

A lot of times, two or three of us had to ride with each other because we couldn't afford to go alone. Cale Yarborough and I went many times where he had the car and I had enough money to buy the gas, and we'd go. He'd say, "Since I own the car, you have to drive."

I've always thought that was so important to becoming the type of driver I ended up being. Had it all been given to me, I don't think I would have enjoyed it as much when I did start winning.

The first job I had outside of racing was in construction, and I made, before taxes, $48.50 a week. A dollar then was like $20 now, though. I worked other jobs, too. I worked at a glass company for a while. Then I started selling cars at Martin Guy Motors, which was a Chrysler dealership in Charlotte.

I could always tell them I needed time off to go racing. There were many times I would leave for a 7 o'clock race in Columbia at 4:30 in the afternoon. That was a race within itself, just getting there. You didn't have interstate highways back then; there weren't even many four-lane highways. There were places where the police just started to wave at you as you went by. They'd stopped you so many times that they felt sorry for you. You'd be in the wind, and the policeman would see you and just throw up his hand at you.

I would get home at two or three in the morning and be at work by 7:30. But it was worth it to me. Getting a lot of sleep doesn't mean you're getting a lot of rest. When I went to bed, I went to sleep. I didn't lay there and worry about the next day.

Later on, after I was married and we had a son, I was selling cars and making a little bit of money, but that's when I finally was told that I had to become either a car salesman or a race car driver. I couldn't do both.

Weekends for a car salesman were pretty big. I had a knack for selling and could have moved up in the dealership, but I would have had to give up racing. They made a decision for me that I needed to make. Bill Taylor was the guy's name at Martin Guy Motors. He said, "You've got a choice to make. It's not fair to the other guys who have to work every weekend while you're off racing."

I decided I was going to give racing a try. That was the difference; when I had to do it, I started racing more—racing five nights a week instead of two or three—to put bread on the table.

It wasn't much of a choice. Nothing meant more to me than my profession, and that was racing. I absolutely loved what I did. If you love something the way I loved racing, everything I did in my life I was flat out and half-turned over.

Now that I look back on it, that might have been the best time in my life, because the pressure then was not what it was later on when I was driving for the factory teams.

I just loved it. Somebody would ask me how much I made that week, and I didn't know. I didn't care. I was paying my bills and having a good time. It wasn't about what you got paid; it never was.

I never read an entry blank, even later on. When I won the Daytona 500, I had no idea how much it paid to win. I told them it didn't matter. Then they said, "You won $104,000." And I said, "Well, now, that matters."

I wonder if I intended to make my hair look like that in this photo with Dad?

I would look down at the bottom of the entry blank sometimes to see what you were guaranteed for making the race, because a lot of times that's what you made.

I quit racing every other day when I first started. After four or five years, I said "Dad blame it, with all the effort I'm putting into this and not getting any more out of it, I'm getting out of it."

But that always went away before it was time to go to the next race.

4

Possum and Pearly Joe

Once Buddy decided he wanted to follow in his father's footsteps and go racing, he started out in late-model stocks, modifieds and just about anything else he could get a ride in on the short tracks around the Carolinas.

One track that played a big part of this developmental portion of Buddy's career was a half-mile of dirt in the South Carolina state capital of Columbia. Built in 1932, Columbia Speedway played host to 43 NASCAR Grand National races —the circuit now known as Winston Cup. But it was the track's weekly racing events that meant the most to Buddy's career. The track closed in 1977.

Columbia Speedway was like tracks in Birmingham or Nashville back in those days. If you could run and keep up and be in the top five regularly at Columbia, you could go other places and whip the tar out of people. They ruined it when they paved it in 1971, but when it was a half-mile dirt track it was something special. It was unbelievable.

A typical Thursday night lineup down there would have Cale Yarborough, LeeRoy Yarbrough, Tiny Lund, David Pearson, Bobby Isaac, Ralph Earnhardt, and people with names that sounded like comic book characters—Possum Jones— who were great drivers. We called this one guy "Pearly Joe" because of his smile. Those guys were great, and if you could run with them down there, you knew you could do it at other places.

I saw Curtis Turner drive in a business suit there one night. He got there too late to change clothes. He never took his tie off, and he won the race.

Columbia was always a sellout, and the promoters were about as colorful as the people driving. Every time you went there you knew you were going to see some great racing.

One night down there I saw LeeRoy Yarbrough get mad at Tiny Lund and dive through the air at Tiny. LeeRoy hit Tiny square in the face and knocked his own self down. LeeRoy went backward when he hit Tiny, who was six-foot-six and every bit of 300 pounds.

Tiny reached down and grabbed LeeRoy and rared back to swing at him. A policeman was there, and he threw his leg out to block the punch. It broke the policeman's ankle. That's how hard Tiny was swinging at him.

Later on, I told LeeRoy, "I know you are smart, but you're not acting smart. Leave Tiny alone. That's a dangerous man."

I could just about do a book on that place alone.

Ralph Earnhardt used to run up behind people on the last lap and touch you just enough to move you. You wouldn't even feel it sometimes. You'd be wondering how you overshot the corner and it would have been Ralph moving you over just enough to get you out in the berm, the area of dirt built up on the outside of the racing groove. You would finish second, he would leave that for you, but Ralph would win.

There was one guy—and I won't say who it was because he's still around—Ralph moved him so many times in one year that on one night, he was leading Ralph on the last lap, and Ralph had a problem going down the back straightaway. This guy went in the corner expecting to get moved aside by Ralph and just spun out on his own. Ralph was stopped halfway down the backstretch, but this guy created his own problem waiting on Ralph to get him.

Certain race tracks just have a certain atmosphere. What makes Birmingham or Nashville special? We had to have to somewhere to come from. We raced at Spartanburg during the fair there and at Greenville, Savannah, and even Jacksonville every once in a while. In North Carolina, we'd race Concord and places like that.

I saw Bud Moore get so mad at Concord one night, he picked up a rock—well, it was more like a boulder than a rock; it was so big he had to pick it up with both hands.

Tiny Lund had put Joe Weatherly out when Joe had about a two-lap lead in one of Bud's cars. Bud got so mad that when Tiny came down the front straightaway, he chunked that rock at Tiny's car. If it had hit him it would have knocked Tiny out the back window. But Bud couldn't throw it far enough. They had to put out a caution to get the boulder out of the way.

5

Hit and Run

You can't be around racing for 61 years like Buddy has and not run across an array of colorful characters and go through some wild adventures. Here are a few short scenes from the early years of Buddy's career.

There was this guy named Marion Cox. He was a legendary car owner in the modifieds throughout the Southeast. Everybody called him "Preacher." Cale Yarborough drove for him; so did LeeRoy Yarbrough. Anybody who has ever been anybody drove for Preacher at one time or another.

We called him Preacher because he would not race on Sunday. He wouldn't even tow his car home on Sunday. If he was far enough away where he raced on a Saturday night that he couldn't get back to Marion, S.C., he would park the car until Monday morning.

I drove for a guy named Spook Crawford. They pulled the race car with an old taxicab DeSoto, and it was as long as two blocks. Spook was a very, very little guy, but his wife must have been close to 240 pounds, and she was meaner than a snake.

I ran the first World 600 at Charlotte for Spook in 1960, which is probably the strangest race I ever ran in. They were late getting the track finished, and when we got there it wasn't nearly ready to take the punishment of 600 miles from those big old cars.

We had to install stuff on that race car that made it look like one of the cars from Mel Gibson's tour through the desert in *Mad Max*. It was incredible.

We put mud flaps on them because big chunks of the track were coming up. I don't know why they even bothered to pave it. It would have been better off if they'd left it dirt. It would get these huge holes in the pavement when you would run on it. When they built Charlotte, the whole infield was nothing but one big rock, and the gravel would fly up out of those holes and just about take your head off.

We knew it would be bad before the race started, so they let us start modifying the cars. And this was back when you still had to leave the door handles on the car. They had to be stock appearing. We put wire across the windshield and put rock deflectors on the front, like cow-catchers on the front of a locomotive.

Jack Smith had a 14-lap lead. That's how bad it got. Then a rock ruptured his gas tank. They put everything in there they could think of to stop that thing from leaking—even a condom, which they blew up and tried to tie it off. They tried everything. Finally, somebody thought about trying Octagon soap to seal it. It didn't fix it either, though, and Joe Lee Johnson won the race.

I finally launched my engine right there in front of the grandstand. It put oil and water everywhere, but it didn't hurt anything. There was no grass there, so I just pulled over in the red dirt.

When I look at Charlotte these days, it's like walking into a futuristic movie. I remember when 30,000 was a big crowd there. Now they have 30,000 in the garage area on race morning.

This is me in my best "James Dean" pose.

I drove for Fred Harb for a while. He was one of the nicest men I ever met and a pretty decent driver. I would show up and he'd have a car there. He'd say, "You want to run next week?" I'd say, "That would be neat." He'd say, "Okay, show up."

I did that a lot. I'd go to some place like Atlanta, and if somebody looked a little spooky on the race track, I would walk up to the car owner and say, "I can do better than that."

They'd say, "Really? Take it out and let's see."

I rode to Daytona with Darel Dieringer one time. He told me we would make it in six hours and something. I told him that was impossible. It was at least an eight-hour trip on a two-lane highway, back before the interstates.

He wasn't kidding. When we rode up he said, "Look at your watch," and it was six hours and a little bit.

I got out and looked at the tires. Where "GOODYEAR" started on the sidewall, there was part of the "G" missing and parts of the "Os." He was leaning the tires under to the point it was wearing the sidewall out.

I said, "I couldn't believe you could make that trip in six hours, but after riding with you I can. And I don't want to ride home with you."

The first race I ever really led much was for Bernard Alvarez in the No. 10 car at North Wilkesboro, N.C., in 1964.

He was one of those guys who was 20 years ahead of himself. The car would fly. I enjoyed driving for him.

I probably could have won that race. The car was that good. But he had a Fiberglas gas tank in it. It looked just like metal, and he had somehow sprayed it with something so a magnet would stick to it. But I went down into the corner, and two or three cars were wrecking. One of them just barely touched me, and it just split that tank wide open.

———————————

Cale Yarborough and I sort of took turns driving the car that Jabe Thomas owned.

If you drove for Jabe, you knew that what you had in that race car was what you were going to run.

Cale drove it one night and told Jabe, "The steering on this thing is so slow, when I go down in the corner my arms go all the way over all crossed up."

Jabe said, "That's easy to fix; just start out with them crossed up and uncross them when you need to."

He didn't have a lot of money to spend. A lot of people just didn't have the means to run any better. A lot of people were good drivers, but they couldn't afford to run any better than they did.

I am glad I had that early in my racing career. Some people never drove anything but good stuff, but I knew what bad stuff was, and I knew how to make it go good enough to get into the position where the factories started helping me.

———————————

I remember when they came out with the first shoulder harness.

I said, "I am not going to strap myself into a car where I can't bend over if I see something coming through the windshield." Then I drove a race in one and knew I'd never even think about getting back in a car without it.

I was racing a late-model stock car at Columbia one time and brought it home and washed all of the red dirt out of the inside of it. The seat belt was an old tank belt made out of cotton, and I left it out in the sun to dry out.

I went back to Columbia the next week and was going down the back straightaway when the belt just came apart. I looked down, and it had rotted and fallen off. Being young and stupid, I put my left leg around the steering column and finished the race thinking that would help if I hit something.

I told Dad about it, and he said, "I ought to fire you right now. Anybody who doesn't pull in when the seat belt comes apart, I am not sure you are going to make it. If it ever happens again, you find a way to come in."

It's hard to believe I was ever this skinny.

I was just out of high school and got married to Colleen Estes. We met in junior high school, and her brother played football with me in high school. We dated all the way through high school, and I was already racing.

It's very hard to put a person in the situation she was put in. When you're first starting out in racing, your picture is the thing they see most in the house. Most of the time you're gone.

I remember going to Maryville, Tenn., one night after our son Bryan had just been born. He was wrapped up in a blanket, riding with us in this old Chrysler, towing the race car. We had the tools in the trunk of the car and our baggage behind us in the seat. The car itself was heavy enough to blow out the brakes, and with that race car pushing it behind, we went over Black Mountain. By about the third turn I went through, I had no brakes at all.

I told Colleen she'd better brace herself pretty good because I didn't know what was going to happen. It was three miles to the bottom of the mountain. There were times I could actually feel the tires far enough underneath that I could feel the rims touching the road.

By the grace of God and plenty of luck, we made it. It went up the next hill, and I just let it choke itself out almost. I pulled off the side of the road and was just sitting there.

"What's wrong?" she said.

"What do you mean?" I said.

"Why did we stop?" she said.

"We just came down the mountain with no brakes," I said.

She looked over at me and said, "I wasn't worried about you making a mistake."

I said, "Well, that's one of us."

The first time I went down to Daytona and had a chance to race, my father wouldn't let me run. He said I wasn't ready and said the car I was supposed to drive was very dangerous.

"If you want to kill yourself, I've got a pistol in the car," he said. "Just turn it on yourself. At least you'll know what killed you."

About five laps into the race, as I was working in the pits as jack man on Dad's car, I saw the car I was going to drive turning end over end down the back straight. The driver was hurt bad.

"I didn't have any place to go." That's me in the No. 87 hitting Doug Cooper's car during a race at North Wilkesboro.

After the race, Dad said, "You see what I am talking about?" He could see what was going on.

If I had been in it, I might not have had many racing stories to tell.

———————

In 1965, I was driving for my father, and I met a guy named Bill Holman. He was one of the hardest working mechanics I ever ran across; he wound up working for Bobby Allison. He had been a race car driver, but he was one of those guys who could take a piece of metal and make a fiddle out of it. He was just very creative, and he understood me.

When he first went to work for us, I told him, "I need somebody to make me a star." When we went to

Me racing on the dirt somewhere. Notice all the elaborate safety equipment we had.

Daytona that year, I got to my hotel one night and there was a big old star hanging on the door to my room.

We bought a car from Cotton Owens that David Pearson had worn out for about three years before that. It was a candidate for the junkyard. We called it the Old Goat. We did incredible things with that car. It was a Dodge, No. 88, and we had some top-five runs that you couldn't ever dream of.

We almost had a big explosion at the shop because of that car.

We were cutting the quarter panels off to reskin it. I hollered, "Whoa! Whoa!" I saw something shiny in there. There was the nicest little five-gallon tank above that left rear wheel, and we were just about to stick a torch to it. It would have lifted our little shop right off the ground if I had. That was where they'd hidden the extra gas to get around the rule you could only have 22 gallons in the tank.

That was the first really good race car I'd ever had.

Sometimes you just have something special with somebody, and Bill Holman and I had it. That's when I became sure that I was good enough to win. I could see I had the talent if I had people like him working with me. That gave me the confidence I needed to go on.

That car was so good that Dodge started sending me parts. They sent me a truckload of engine parts and body panels.

The guy just pulled up to the shop and started unloading the truck.

My dad told them, "We didn't order anything."

I got on the telephone, called the Dodge people, and said, "What's going on here? We can't afford this stuff."

They said, "Buddy, it's a gift from Dodge."

I said, "That's mine?" It was more stuff than we'd ever had in the shop. We didn't know how to act. A new transmission?

Most of the stuff I'd run before that, I had bought from somebody that had already wore it out, just like the "Old Goat" was.

6

Getting Even With Fireball

One of Buddy's real heroes among drivers was Glenn "Fireball" Roberts, one of NASCAR's first superstars, who won 33 races before dying of complications from burns he suffered in a crash in the 1964 World 600 at Charlotte Motor Speedway.

Fireball Roberts was one of the first real friends I made. He was kind of "the man" at the major speedways. He was the first person about whom I said, "If I could ever get to where I could run close to him, that would be the biggest thrill in my life."

Daytona would eventually come to be one of my best tracks. When I first got there, though, I would always show up with some little old car so uncertain that you didn't know whether the tires were going to rub or what. So I would go down the back straightaway on the first or second lap, wiggling it back and forth to make sure the tires didn't smoke and looking at the gauges. It was like reading a book; you wanted to know the contents.

I would get down the backstretch, and Fireball would come out of nowhere. He would go flying by me and miss me by about an inch and a half, and I would almost spin the car out. It would scare the living tar out of me.

When I'd been there about four or five years, Smokey Yunick had built a brand new car for Fireball. I'd already been out before he got up to speed. He was going down the back striaghtaway doing the same thing I'd been doing, looking down at the gauges and checking it out.

I came by him, "Whoosh!" I mean I almost took the numbers off that thing. I saw him; he grabbed that steering wheel and almost wrecked it.

When I got in, he was already puffed up.

He wiggled his finger at me and said, "Come here."

"What?" I said"

"Let me tell you something," he said. "That wasn't even funny. I'm good enough to do that. You're not."

7

Two Lead Dogs

Cale Yarborough, like Buddy, had only one setting—wide open and hammer down. Yarborough came from Timmonsville, S.C., just up the road from Darlington Raceway, and remembers sneaking through the fences there as a boy to see the races. Later, of course, he would become one of the greatest champions in NASCAR history and one of Buddy's best pals.

Cale was one of my best friends, but on the track we were like a piece of metal and a magnet.

We crashed one time at Martinsville before the green flag ever came out. He was on the pole, and I was on the outside pole. We started rubbing coming out of Turn 4 on the last pace lap, and we never made it through the first corner. We just went straight on into the wall, but it started way before the green ever flew.

I guess we just raced too much alike as far as deter-mination goes. We had respect for each other, but some-times it just didn't work out.

One time at Bristol, I was leading on the white flag lap, and I heard something coming across the apron with the motor running wide open. He got me up into the wall, and that could have been all there was to it. But I held it wide open, and when I came off the wall I got him in the right-rear corner and he went up against the wall.

He was lucky enough to spin backward faster than I was going around. He had a little better slide angle than I did, and we finished first and second. But I wanted to make sure he was spinning, too.

That went on for about five or six races. We just wouldn't let it die.

We went to Pocono for the next race. I think he was running second and I was running third. We started into the third turn, and I never turned. He got out in the loose stuff and I got the spot. It got to a point where it was getting serious. We weren't just playing on the short tracks.

We got to Darlington, and as it would happen, we walked through the infield gate together. We almost bumped each other going in.

I said, "You know something, Cale, this crap has got to stop."

He said, "All I can say is get your best hold."

When he said that, I went straight up. I said a few things and he just kept walking, thank goodness.

We didn't have much trouble after that. We were just too much alike, wanting to lead every lap.

Cale Yarborough (center) was a close friend off the track, but somehow we always seemed to get in each other's way when we were racing together—probably because we both like the view from up front so much. That's me on the right with legendary racing announcer Ray Melton on the left in a photo taken in 1965 in, I think, Norfolk, Virginia.

We both knew that old expression: "If you're not the lead dog, the scenery never changes."

One of the funniest things I remember is a time when Cale and I went to Nashville. Neither of us was a really big name; we were still running late models around the country. We had made enough of a name that we had some value to the race track, and they called us to come up there.

The promoter up there promised us $500 over what we made. Cale and I ran in the top five, and both of us made more than $500 each from the purse. We went up there to get paid, and the guy said, "I didn't promise you $500 over what you made; I promised you'd each make at least $500."

That is when I found out how stubborn Cale could be. He wanted his money, and he told that guy, "Listen, we talked, and I told you that if I won the race I still wanted my $500. And you said yes."

The promoter said, "I'm giving you $100 and that's that." He handed each of us a $100 bill.

Cale wadded up that $100 bill and hit him right in the head with it. And when he did, the promoter reached over and pulled out a pistol. He said, "This has gone as far as this is going to go."

Cale said, "I'll tell you one thing. I am not taking that money, but you will pay for it someday. Every time I'm ever going to come up here, I will get that back ten times over." And every time he was invited back, Cale would make him send the money ahead of time.

Somehow, though, I ended up with Cale's $100. I guess he had more principle than I did.

Cale is one of the few I know who could drive anything. People have told me they would take absolute junk to the track for him to drive, and Cale would go out and find a way to make it win.

He was my archrival, but there's no race car driver who drove harder than he did with 10 laps to go. Nobody. I raced with three generations of drivers, and there was never anybody more determined than Cale. He was made to race. He had no neck and was as strong as a horse. He understood what it was all about, and you could never count him out. He could take a bad car and find someplace to make it work on the track, and he'd wear you out if you weren't careful.

8

Here's Mud in Your Eye

If there's one Buddy Baker story that's told more often than any other, it's probably this one about his misadventures on one night of racing at Smoky Mountain Speedway in Maryville, Tenn.

Buddy says it's the favorite story of Don Naman,[1] who's now the executive director of the International Motorsports Hall of Fame in Talladega, Ala. "He tells the story better than I do," Buddy says, "because there were moments there I was about to black out."

We got talked into bringing one of Ray Fox's Dodges, one we ran at major speedways, not a short-track car, up there to Maryville, Tenn., to run at this track Don Naman was running.

We went up there, and I sat on the pole by 400 miles. It was a half-mile track banked like Bristol, and that suited me just fine, but the tires were good for about 15 laps the

---

1. Editor's Note: Naman passed away in 2011.

way I was driving it. We started the race, and I got way out in the lead. I mean, I was gone.

I said, "Well, this is going to be fun here."

I went down into Turn 3 and felt this little rumble. I thought a wheel weight had come off or something.

I went down the front straightaway, and when I turned there was a "Boom!" The tire blew.

I went straight in, head first, and hit the wall like a ton of bricks. It knocked me crazy and I slumped down, and then I looked over and saw this big Indian head coming at me. The ambulance they had that night was an old Pontiac hearse, and the emblem, the Indian head, on the front lit up red when the lights were on. I saw that Indian head go by the door, and these two guys got out, Bubba and some guy who looked like Barney Fife.

First they tried to pull me out of the car by my head. I said, "You've got to unhook something first."

I knew I had broken ribs, there was no question about it. They finally pulled me out and put me on this old gurney with casters on the bottom in the back of this hearse. I noticed they didn't lock the wheels or lock the back door.

We went down the back straightaway, and the straightaway was banked almost as steep as the turn. They opened the gate for us to go out, but they hadn't red-flagged the race. They just had a yellow out. A bunch of the cars had been in the pits, and we started to make our turn to go across the track. The ambulance driver looked and saw all of these cars coming off pit road, and he nailed it wide open.

He got about halfway across the track when the back door flew open, and here I came out.

Now I'm on the gurney with my arms and legs strapped down, and these cars are coming at me at 120 mph. I got one arm out and started waving my hand at them a little, because back then on those tracks the lights were horrible. I was going down the hill, and they saw me. One car went down through the mud and the other one went by on the high side. Two drivers told me later they would have hit me if I hadn't waved at them.

But that wasn't the bad part.

I got down to the bottom of the track, and that gurney was running about 30 mph when it hit the mud on the inside of the track. Those little wheels burrowed in, and it went straight upside down. I went down in the mud to my ears on both sides. One of the guys from the ambulance jumped out, grabbed me and rolled me over.

He said, "Are you okay?"

I said, "If I ever get off this thing, I am going to kill you first."

9

Victory—At Last

It took Buddy until 1967 to finally put everything together and score his first Winston Cup race victory. Fittingly,
it came at his hometown track, Charlotte Motor Speedway,
on Oct. 15 in the National 500.

Buddy finished a lap ahead of the rest of the field in a
No. 3 Dodge owned by Ray Fox. He won $18,950—an
amount that seemed like a princely sum at the time.

More importantly, however, he won the approval of the
man he loved and respected more than anybody else in the
world, a tough old former Winston Cup champion who came
to the winner's circle that day to tell his son that he'd done a
good job.

That was the day when the frog felt like a prince.

I spent a long time building up to winning in Winston Cup. From 1959 to 1967 is a pretty good little
stretch. I'd run in more than 200 races before I won. I'd

already had a son, and I was bound and determined I was not going to fail.

You'd tell yourself that you were a better driver than people who were winning. I knew I was.

Ray Fox invited me to drive his car. I borrowed the money from my mother and rode a bus to Daytona. This was in the winter before the 1966 season.

I went to Ray's shop and helped fix up the seat in the race car. But one of the people involved with Dodge kind of didn't like the idea of putting me in the car. He didn't think they needed two drivers in the Fox group. So, as I was putting the last bolts in the seat and seeing visions of winning my first race dance in my head, they called and said it wasn't going to happen.

You want to talk about a long bus ride back home? That was it. I came back and did the best I could for a while.

Ray called me back and said things had changed and he again wanted me to come to drive for him. I thought about it and decided I just couldn't stand another disappointment. I couldn't bear having my heart broken like that again.

At first, I told him I couldn't give him an answer right off. I thought about it some more, though, and said to myself, "Are you crazy? This is everything you've been wanting to do."

So I went to Charlotte for the World 600 in that car. I had done all of this bragging and talking about showing people what I could do if I ever got a shot.

I got to the gate to go into the infield, and it was as if my car had a reverse hooked up to it. I turned around

A big hug from Mom after my first win at Charlotte in 1967.

and went up on the hill that kind of overlooked the track and sat there for a few minutes.

I went back to the gate and stopped again.

"What if you can't do it?" I asked myself. "You've been talking the big talk. What if you can't do it?"

So I went back up on the hill again.

The third time, I thought, "By darn, I am going to go in there."

I'd made it up in my mind already. No matter how I did, I could say I was just getting used to the car. I took off. I went out there, and that thing was a bullet. I drove it as hard as I could possibly drive it.

Finally! This is me celebrating my first career victory in NASCAR's top series at my hometown track in Charlotte in 1967.

I came in to the pits, and everybody was looking at me.

I thought, "Boy, I must have been really bad."

I said, "Ray, before you say anything, I am just getting used to the car and I will get going better."

He said, "Boy, I hope not. They'll run us out of here. You're about half a second faster than anybody's ever been around this place."

I said, "Cooool!"

When the race started, I was out of there. I was leading, and for once the crowd was cheering me for that instead of for me coming up from 29th or something like that to the top five.

From then on, I knew it was a matter of time.

I eventually cooked the motor in that race, and we had tire problems. Just about everybody Firestone had in that race wound up in the wall. They did some testing with me at Charlotte, and we came back in the fall and kicked some grown tail then.

You've heard drivers say they could hear everything in the car. It was like I could hear the oil flowing through the engine. It had been a dream of mine for so long, and we were beating everybody. We had a lap on the field. It was car No. 3. The only thing that could happen would be something like that.

That first time you make the hard left and go to the winner's circle—you talk about not being prepared for something; I had no idea what to say or what to do.

All of a sudden people were calling me "Champ!" and stuff like that.

I was thinking that three years earlier, that same guy had been telling me I should hang it up because I would never make it. The same guy who came up and said, "I knew you would make it! It was just a matter of time." He'd be the same guy who'd been saying, "Buck has to be aggravated by the way he's driving."

You can say what you please, but the first time you win a major race, it's beyond belief. I eventually won the Daytona 500, but even that didn't have the same feeling as that first race.

Dad had broken his leg before that, and he came on his crutches to the winner's circle. He had never, ever, in the whole time I'd been driving, said anything complimentary. It was always, "If you would have done this or if you hadn't done that, you would have run better."

When I won, he just grabbed me by the back of the head and pulled me down. There were tears running down his face.

And he said, "You know, you did almost as well I would have done."

Later on, he told me, "To see you run a major speedway, it's something. I just can't tell you how proud I am." But in the same breath, he'd say, "But you can't rest on that. You need to race like that all of the time."

The day after that race, Bob Latford, a guy a lot of people in racing have forgotten already, helped me put things in perspective.

Bob, one of the guys who developed the points system they use in Winston Cup racing to this day, was doing public relations for the Charlotte track. They had a promotional thing they wanted me to go out on the next day.

I went out there and I said, "Bob, I don't want to do that."

He said, "Buddy, you're right. You won the race. You're a very important guy."

He had a picture with him, one that somebody had taken from a satellite. You still hadn't seen too many of those back in 1967. It showed the whole world.

Bob said, "Yeah, you're very important. Find yourself on there."

I did the promotional tour, and I tried to remember that my whole career.

The winner's interview after the 1967 National 500. That's Bob Latford on the left with the microphone and my car owner, Ray Fox, in the white shirt beside me.

10

The Quiet Men

H. A. "Humpy" Wheeler, the longtime president of what is now Lowe's Motor Speedway in Buddy's hometown of Charlotte, once led the development program for Firestone's racing tires. Among the drivers he employed to push those tires to their limits was Baker, a job that Buddy says paid dividends in many ways throughout his career.

Wheeler inherited his nickname from his father, who got it after being forced to run laps by his football coach after getting caught smoking Camel cigarettes. The elder Wheeler eventually moved to the textile town of Belmont, N.C., near Charlotte and served as the athletic director at a Catholic college there. Little Humpy learned to box because he felt like he needed to know how to take care of himself in the rough-and-tumble environment of his youth.

That training would later prove helpful as he provided another kind of assistance in helping toughen Buddy up for competition at NASCAR's top level.

When I went to work for Ray Fox, Humpy Wheeler was working for Firestone. He saw something in me that told him I would make a good tire tester. I guess it was because I didn't have enough sense to back off.

I ended up having the luxury of going to Daytona and spending a week testing tires. I'd go to all the major speedways and test, and that gave me such an advantage. It was like kick-starting me and getting me on the right track. Not only did I get good set-ups, but it helped our race team.

Tire testing for Firestone helped me become the kind of racer I wound up being on the superspeedways. That's H.A. "Humpty" Wheeler, now the president of Lowe's Motor Speedway in Charlotte, second from the right beside me. Humpy helped me in many ways throughout my driving career.

Everybody else was running to Greenville and Columbia and here and there. I was making $10,000 or $15,000 a test, and that was big money back then. A guy winning the race at Columbia was winning maybe $1,000. I thought, "Let's see, $10,000 or $1,000? That's not even a contest."

I did a lot of testing for Firestone, and that helped me a lot on the major speedways. They liked my input, because I could tell them when the tire started to give up and why. Humpy took a personal interest in me and must have felt that I had the need for some special training.

I took it on myself to run every day, and I thought I was pretty fit, but I was nothing like he thought I should be. He told me that if I worked out with him every day for two weeks leading up to a race at Charlotte, he would guarantee that I would win the race.

It has always been like he could see the future. He's wrong sometimes, but a lot of times he's right. He told me that he was a boxer, and a pretty good one, and that if I would work out with him for two weeks he would have me ready. We would ride from his office out to this field in a car with the windows up and the heat on high.

A lot of times in October around the Carolinas it's still pretty warm. It was about 85 degrees and we'd have the heat wide open. We'd get out to the field looking like two seals. Then we'd start working out. We'd run and then box—bang, bang, bang—and man, I was getting as hard as a brick. Every pair of pants I had was about four inches too big around the waist. I was in great shape. The top part of my body was getting bigger, and I was losing weight.

Charlie Glotzbach gets his "Darlington stripe" by rubbing the guardrail in Turn 4 in the 1967 Southern 500. That's me staying off the wall—at least this time around.

We'd done this every day for two weeks, and on the last day before they started practice at Charlotte, for some reason, just was we were going to start working, out he said, "No hitting in the head." We hadn't been doing that for two weeks; we'd been hitting each other with body shots and that. I found it strange that he would say that. I must have hit a nerve the day before, because we'd gotten pretty aggressive.

The first lick he put on me that last day, he almost tore my right ear off. When he did, the trigger in me went off. I thought, "Okay, let's go."

The fur was flying, boy. We were bleeding and really having at it. He hits like a mule kicking anyway. We went across that field, through the ditch, through a man's rose bushes across the road there and up into his carport. It was like that fight in the John Wayne movie, *The Quiet Man.*

About that time, Humpy put both of his gloves up and said, "We'd better stop before somebody gets mad."

I said, "To hell with you, I've been mad since we left that field over there!"

We rode back in the same car with steam coming out our ears. Humpy knew it had gone far enough. We could have become enemies right quick that day.

But you know what? I won the race that weekend.

11

Me and Tiny

One of Buddy's best friends in racing was a giant of man who, in typical racing fashion, carried the nickname of "Tiny."

DeWayne Louis Lund, born in 1936, was from the tiny town of Cross, S.C. Tiny Lund stood six feet, six inches tall and weighed in the neighborhood of 300 pounds. His most famous feat in racing was winning the 1963 Daytona 500 in a car owned by the legendary Wood brothers in what can only be called storybook fashion.

Marvin Panch, the Wood brothers' driver for the 500 that year, had crashed badly in a sports car while trying to set a speed record. Lund was among a group who ran to Panch's aid and pulled him free from the fiery crash. The Wood brothers asked Lund to drive the car because Panch was hurt too badly to compete, and Tiny won the race.

"Tiny was like a big brother to me," Buddy says. "He'd raced for my dad one season, and for some reason we always got along really well."

Well, maybe not always.

We were racing up at Martinsville, and Tiny had been in the pits with some kind of problem with his car. I was having a great race with Richard Petty and, I think, Bobby Allison, and here comes Tiny out of the pits, running 25 laps down.

Those two guys were giving me a fit, and Tiny pulls out right in front of me. I kind of chunked him in the rear end one time to get his attention, and he didn't move,

A group of Playboys? From the left, that's Dick Hutcherson, my good pals Tiny Lund and Cale Yarborough, a Playboy bunny whose name I don't recall—honestly!—Ned Jarrett, me and Buck.

letting Petty and Allison run right up on me. We went down to the next turn, and I went in too hard, lost the front end, and hit Tiny so hard I bent the door bars. It knocked him out of the race.

After it was over, I looked up and here came Tiny down through the pit area. If his feet were an inch long they were size 14, and he was coming at me with those big old gorilla arms of his. Every time his feet hit the ground the dust went flying.

This was before I knew him well at all. Later on, I would have started laughing when I saw him coming. But at that point I wasn't sure what was going to happen.

I said, "Oh, Lord." I looked around; there was an axle in a bucket, and I started to pick it up. Then I thought about it. I said, "Don't do that, he will take it from you and kill you with it."

He came up to me, and my adrenaline was still pumping.

He said, "You wouldn't kill somebody to get by them, would you?"

I don't where it came from, but I said, "Of all the people to say anything, you should be the last one! You've run into me five times in one corner!"

He just looked at me and swung that big old arm around my neck. He put my head up under his arm and buzzed the top of my head with his knuckles and started laughing.

We became great friends, but he liked to set up little things that would make me so mad I couldn't stand it.

In the early days of my career, I hadn't realized that getting to bed by, say, 9:30 would be a good idea.

I had a habit back then of coming in and not wearing pajamas or anything. One night up in Nashville I was out until about 11:30 or so. Tiny was out with us—until about 11:15.

They used to travel and promote the races at Charlotte Motor Speedway with a mascot, and the mascot was a fully grown cheetah. It was waist high. Tiny knew they had this animal there at the hotel.

We'd been out having a good time, and I came in and did my normal thing. I just took my clothes off right after I came in the doorway and just hung them up on the floor. I never turned on the lights; I just went straight to bed. I was really ready to go to sleep, and I just pulled the covers down, got into bed and pulled the covers back up.

Right as I got the covers pulled up around my neck, I heard a noise. "Rrrrrrrrrrr."

I thought, "What in the heck? I know I had dinner, my stomach can't be growling that bad."

I just tried to roll back over and shut my eyes and I heard it again.

"Rrrrrrrrrrr."

I went, "Wait a minute!"

I reached over and fumbled along the edge of the end table until I found the light. I bumped the light on, and when I did, that cheetah was about two inches from my face.

Now, I am the kind of a person that a yard dog kind of scares me. I made a mad dash for the door and ran out in the hallway as naked as the day I was born. I probably scared the cat about as much as it scared me.

I was running down the hall, and the first thing I saw was Tiny Lund. He started running backward—he thought I was coming to visit with him about putting this cat in my room. I thought he went back in the room because the cat was still chasing me, so I kept running on down the hall, making enough noise to wake the dead.

Everybody I met would turn around and run because I was tearing down the hallway, buck naked, screaming like crazy. Every time I saw people looking afraid, I picked up speed because I thought this cat was coming to get me. I turned to go down the steps before I realized I was the only thing running down through there.

I finally went back to my room. You know what a motel room door does when you go out it, don't you? It locks. All of my buddies were there to make sure they didn't offer any help. Finally, I hid down near the drink machines and hid myself until I saw somebody I could send to get a spare key.

We would go fishing together and Tiny would aggravate the living starch out of me. He just couldn't help himself. He loved to pick at me, and I didn't take it very well. I would gripe at him and bellyache.

I would open a sandwich and he'd pick it up and take a bite out of it. I would say, "You eat it now." I'd open a Coca-Cola and sit it down and he'd pick it up and take a sip. He knew I wasn't going to have it then.

I went down to his fishing camp at the Santee-Cooper lake in South Carolina one weekend, and he said, "I don't know what in the world we're going to do about these alligators."

Now he knew that if it was real hot, I was going to go swimming.

He kept saying, "This year, we've seen more alligators in this damn lake. It's to the point where people are afraid to tie up. You just can't believe it."

About 4:30 or 5 o'clock in the afternoon, it was 95 degrees if it was 1 degree, and I said, "Well, I am going swimming."

I stripped down to my shorts and dove off the back of the boat. I went down and was paddling around out there when a pain hit me like you would not believe.

Tiny had jumped off the front of the boat, swam under, and caught me right in the crotch.

The first thing I thought, of course, was, "It's an alligator!"

When you're caught in that spot, you're going with whatever it is. I started swimming trying to keep up. Eventually, I could hear Tiny under there. The bubbles started coming up because he was laughing so hard.

I got back up in that boat, and by gosh, I was about determined that he was going to drown. I wasn't about to let him back in that boat.

I took the paddle, and every time he stuck that big head up I would pop at it. I told him, "I ought to kill you." He had set me up all day long.

There was a saying that if you wanted to find Tiny Lund, you just popped your trunk if you had anything to eat and he would be there. Strong? Oh, goodness, he could have lifted my house, I think.

I won the race during which Tiny got killed, at Talladega in 1975. That was one of the most traumatic things that ever happened to me. I was up in the press box. I had asked my crew, because it looked bad. I said, "How's Tiny?" They never would tell me.

I was up in the press box, and I was happy. I'd won the race, and I was cutting up with all of the press. Some person who didn't know we were friends just said, out of the blue, "Oh, by the way, Tiny was killed today."

I felt like somebody had yanked my heart out through my feet. I never had anything to shock me like that. I said, "Well, this press conference is over."

I went to his funeral, and I saw him there. It just tore me up. That's the way I remember him now, the way I saw him last. I wish I had never looked at him like that.

That's one thing people don't seem to understand. Race car drivers are human just like everybody else. People mean something to one another; there's a closeness between all of us. I didn't have all that many friends among the guys I ran with, but Tiny was just a great guy.

12

The 200 MPH Man

In March of 1970, Buddy became the first man to ever drive a car on an official lap at more than 200 mph on a closed course when he ran 200.447. He did it in a Dodge Daytona prepared by Cotton Owens, for whom Buddy had gone to drive after winning his second straight Charlotte race in the 1968 World 600 in Ray Fox's car.

I was driving a Dodge Charger in 1968, and they gave it the name "America's No. 1 Charger." That didn't sit too well with some of the other guys driving Dodges. But for some reason, Frank Wylie from Dodge gave me that name.

Dodge picked me to run the first official 200 mph lap at Talladega. We had all run 200 mph before, but not officially. When they did pick me to be the one, there was a lot of bellyaching from the rest of the guys driving Dodges. We got a lot of press. It wasn't a big deal for me back then, but it became a big part of my life.

We called NASCAR, and they brought all of the timing devices down there. Even with Bill France standing there looking on, I didn't realize until I saw the numbers on the board that I was the first to officially go over 200 mph.

I could have run 210. We did it in the second or third lap.

That whole bunch from Dodge had their game faces on. The Dodges were unbelievable, with those big wings on the back of them. That car was so good that if you drove it two or three years you would have to learn how

Cotton Owens and I won the 1970 Southern 500 at Darlington on the same weekend Cotton was inducted into the Stock Car Racing Hall of Fame.

to drive all over again when you got out of it. It would turn sideways, and that big wing would straighten it up like the wing on an airplane.

I ran the lap in a car owned by Cotton Owens. My father had won the Southern 500 three times, and I can remember as a kid sitting there underneath the flag stand watching him celebrate in the winner's circle and me dreaming of winning a race there one day.

Darlington always seemed huge to me, but there probably weren't 35,000 there that day. I thought everybody in the state of South Carolina was there, though.

That fall, the National Motorsports Press Association inducted Cotton Owens into the Stock Car Racing Hall of Fame there at the Darlington track on the same weekend of the Southern 500. None of Cotton's cars, though, had ever won at Darlington.

He was a great guy. When you drove for Cotton Owens, you ran for Dot, his wife, too. His son was on the crew, and the boys who worked on the car were like family.

We went through a spell of bad luck when I first went to drive for him. We would lead just about every race, and then something would happen to the car. They gave me a little trophy that said,†"You're our horse if we never win a race." That was when things seemed to kind of turn around.

In the 1970 Southern 500, Cale Yarborough and I went at it all day. We'd swap the lead back and forth, and we were lapping the field like it was backing up.

We had such a battle between the two of us, though, two drivers at the top of their game in two great race cars. We went at it. We ran up on another car, and Cale went to the inside and lost control. He got up on top of the fence, and he must have gone a half-mile with that fence just shearing everything out from underneath the car.

From then on, I was on cruise control. We finished on a lap by ourselves.

Those kinds of races are much tougher than the ones where you're fighting right down to the end. When you're racing and going at it with somebody, they're not tough.

On the white flag lap, I was just cutting up. I came off the fourth turn and smoked the tires, just letting the rear end jump out a little bit.

Cotton called me over the radio.

"I swear to God," he said, "I will kill you if you wreck that car."

Very few people have a winner's circle smile, but he had it that day. You could tell there was a lot of pride in that. He had run so very well himself there. Ralph Earnhardt drove for him at Darlington. But he'd never won it, not even with Pearson and all the people he'd had driving for him. Not until that day.

13

Don't Mess with Texas

Things didn't always go well for Buddy on the track, of course.

The biggest embarrassment of my racing career came at the Texas World Speedway in College Station in the Cotton Owens car. There are some things a driver doesn't like to talk about, but since this is a book about what happened, I will talk about it.

I went to Texas, and that track suited my driving style almost perfectly. I looked at it and thought I would do well there.

I had a two-lap lead on a two-mile track. I was four miles ahead of everybody. My radio had gone out, and the pit crew was talking to me with signs. With about four or five laps to go, I was riding along under the yellow and trying to read this sign they were holding up.

The sign said "You've got it made!" and had a big dollar sign on it.

At that moment, James Hylton's car blew up in front of me. I was reading a sign and ran right into his rear end. We had it won, and I messed up.

We got on that airplane to come home that night, and you could have heard a mouse pee in cotton. It was the quietest flight in history.

I went down the next day to Cotton's shop and said, "Cotton, just go ahead and do what you've got to do. Nobody can be that stupid. I am embarrassed for you and for me, and I would completely understand if we terminated our deal right there."

He said, "Are you crazy? We were the best team there. You made a mistake, and people make mistakes."

Cotton was okay with it, but of course people couldn't stop talking about it. And one of the guys who gave me the most grief about it was Donnie Allison. He made the biggest thing about that after the race, just chirping about it.

All I could say was "Yeah, it was stupid." I couldn't take that lap back or anything. It sure wasn't something I'd planned.

I got to talking to A. J. Foyt about it, and he told me he'd done it once in an Indy car. I started to find out that I wasn't the only person who'd ever done something like that.

Later on that year, Donnie Allison had the field covered at Daytona. He came past the start-finish line and caught up to a bunch of lapped cars behind the pace car. He put his brakes on, and the brakes locked up and he went straight into the back of one of them. When I came by him I couldn't help it. I mashed the clutch in and revved the engine so he'd see me.

I came back later, in 1972, and won at Texas in Nord Krauskopf's car.

We had the field lapped in that race, but they had a big wreck on the frontstretch, and Foyt just came down there wide open and went right through the smoke. You couldn't see anything, but he made it out the other side and got back on the lead lap. Then I had to fight him. The last two laps, I think we changed the lead eight or nine times. I was on the inside and he was on the outside, and I finally moved him over to where he kind of grazed the wall just enough to rub off enough speed that I could get ahead of him.

It wasn't anything mean; you wouldn't do that to A. J. anyhow unless you had roller skates to get out of there on. Not in Texas.

14

Petty Differences

By the start of the 1971 season, Buddy had won three races and established himself as one of the stars in the Chrysler racing family. The factory decided to downsize its NASCAR operations, however, choosing to run only a two-car operation out of the Petty Enterprises team in Level Cross, N.C.

Richard Petty was one of the team's drivers, and Buddy was made the second, bringing to an end his happy association with Cotton Owens's team and putting him into the middle of a situation he knew might be somewhat ticklish.

I was happy as could be with Cotton Owens. But Dodge just decided to downsize and get rid of everybody but Richard Petty and me.

I was going in behind a guy who did very well at Petty Enterprises and built a great relationship with that team— Pete Hamilton, who wound up going to take my old ride with Cotton's team.

Pete did everything you could ask anybody to do in those cars. And I had to go from a bunch that loved me to death and where I had a happy home into that new environment.

I still remember walking into Petty Enterprises the first day. A guy hollered, "Whoa! Everybody stop working! The Messiah is here!"

I said, "Oh, no! What am I into?"

I started going up there just as much as I could. I bought a motorcycle and went up there and rode it around with them, just to build a better relationship.

They liked Pete Hamilton—I liked Pete. It was very uncomfortable, because Pete had done a lot for them. It was tough to go in there. I had been one of the people they'd been battling every week.

Richard Petty, now, was the nicest person on Earth, and he was always there when I needed to talk to somebody.

These teams, a lot of times, are like family, and I had a nice family over there with Cotton Owens. But all of a sudden that was no longer there. Had it not been for Richard and Dale Inman—Dale was great, he would hide and jump on me when I walked in and cut up with me— had it not been for those two, the first couple of races with Petty Enterprises would have been impossible.

We went to Daytona and finished first and second. I caught him with a few laps to go, and I'd been told by the men in charge that first and second was as good as we could do.

They didn't tell me not to race him, but they did make it clear that if I wrecked him I wouldn't have to

Richard Petty and I getting ready to start on the front row at Darlington in 1967. I would later be Richard's teammate at Petty Enterprises. Those are some pretty sporty driving shoes I've got on there, aren't they?

come to Randleman the next week to find out if I still had a job.

As it ended up, we all got along quite well. Just being around Richard was a treat. I was almost doomed when I first walked in, but when we overcame that and started racing together and racing hard it worked.

We won at Darlington in the spring. We won that race by seven laps. Dick Brooks finished second, and after the race he said, "When you lapped me that last time, you will never know how close you came to getting it. I wanted to turn right so bad." Then we won the World 600 at Charlotte the next year.

The guys at the Petty shop set me up good with Lee Petty. Right when I got there, they all told me that Lee was practically deaf, that he couldn't hear a thing.

LeeRoy Yarbrough and I got into it at Rockingham, and I hit him, knocked him up out of the way, and went on. Richard and I both finished in the top five, but Lee didn't say a thing about how well we'd run.

He came walking over the shop, and I was up under the car working.

He leaned over and said, "You'd better be glad I wasn't in that 98 car yesterday."

I said, "What are you talking about?"

He said, "If you'd ever hit me like you hit LeeRoy, I would have followed you home to get even."

He turned around to walk off. I thought he couldn't hear, and I mumbled, "Yeah, you and whose army?"

He was way out across the shop, and he turned around and said, "And you will lose your job if you start mouthing off to me."

I thought, "Oh, no, I've been set up again."

I raced against Lee. We were having a heck of a battle up at the old speedway in Asheville one time, and we got down to it near the end. He just put me right in the wall.

That made me hot. I walked down there and said, "Lee, what in the world was that? There was plenty of race track."

He said, "I don't know. When I turned left, the front end started chattering and it was either back off and lose the spot or stay in it and hope you turned me. You got me turned and I appreciate that."

We got to talking, and I ended up apologizing to him for even being out there.

He and my father had some great races against each other. Lee, my father, Junior Johnson, and those guys were the people who made racing. You should see some of the stuff they ran fast in; it would scare you.

Toward the end of my career I was racing in the July race at Daytona, driving a pea-green car sponsored by Publix grocery stores. I wasn't too thrilled with the color of that car, but I was in it, and I will bet you I was going close to 200 mph going into Turn 3. I had a head of steam up and went to the outside just clicking it off right around the top. I was passing cars like they were stopped.

I came off the fourth turn and went into the front straightaway, and it looked like somebody had opened the gates to the stock yard and the cows were running through the streets. That's what the cars that were wrecking looked like. Ernie Irvan and I made contact once, then went way out away from each other and came back and hit each other again. Then we missed a couple of cars and hit again. The fourth time I saw we were almost going to hit head on, and I just put on the brakes and hit somebody else to get him off the hook. I was just glad I hit something other than Ernie's yellow car.

Anyway, I looked up, and Richard Petty was sitting turned around on the backstretch. I ran across the track to see if he was okay. His car was just done in. He was sitting there with his head down. The windshield had been knocked out of his car, and I jumped up on the front of it. I was looking in and reaching for him when he finally raised his head.

He said, "What?" and pulled back—I don't know if he thought I was mad or just didn't know what was going on for a minute.

I said, "Are you all right?"

He said, "Yeah, I guess."

I said, "Well, get out of there then."

15

Days of Thunder

Buddy split time in 1972 driving for the Pettys and for a team owned by Nord Krauskopf, where his crew chief was a NASCAR legend named Harry Hyde. Buddy would win the 1973 Coca-Cola 600 at Charlotte for that team, and later that year he also got his first career short-track Winston Cup victory at Nashville.

Hyde and his later association with Tim Richmond provided the basis for the character portrayed by Robert Duvall in the movie Days of Thunder *starring Tom Cruise. Some of the things that happened in that film happened in real life, too.*

The most fun I ever had driving a race car was with Nord Krauskopf and Harry Hyde in their No. 71 car. Harry told me right after I went to work for him, "I didn't know how good I was until I got you."

We were running up at Martinsville, Va., and I had come from the back to the front three times. That car literally looked somebody had taken a carving knife to it. There wasn't a straight piece on it.

We started up front, and I had built up a pretty good lead. They had a caution, and I said, "I need four tires." Harry said, "I don't think you do." I said, "I need four tires." Harry said "Come on and get 'em then."

I came in, and something happened in the pits. I went out dead last, and Harry was ranting. You just have not been cussed out until Harry Hyde has cussed you out. He was a professional at it. If he said he wanted to talk to you, if he left the door open it was okay. If he shut

Richard Petty and I racing door to door. I'm in the No. 71 Dodge owned by Nord Krauskopf.

that door, you were going to get a man's cussing.

Anyway, when they dropped the green flag, I started coming up through there. I was going through the middle and on the outside, jumping the curb and running into people—it was pretty ugly.

I got back in the lead, and the yellow came out again. I said, "I need four tires." Harry said, "Absolutely not." I said, "I need four tires." Harry said, "Well, come on and get them." Something happened again, and we went back out again in the back.

This time, I put a "71" on everybody there. Everybody looked like they had our paint scheme. Sure enough, the caution came out again.

"I need four tires," I said. "Absolutely not, no sir," Harry said. "Why not?" I asked.

Harry said, "I want you, when that pace car comes out there, I want you to nail him and just take him right in the wall."

"What in the world are you talking about?" I said.

He said, "That's the only thing going left today you haven't run over. I want you to have 100 percent."

I got on him one time at Talladega. I mean, we had the fastest car down there but I said, "Harry, I need more spoiler. This thing's loose." It was loose, too; it was spooky loose.

He said, "Dad-gummit, you couldn't throw this car out of the race track. But all right. I am going to go make some changes, but I ain't going to like it."

He went back there, working away. He said, "All right, go on out." But he was mad. He was red as a firecracker.

I went out and came off Turn 4 and that thing just jumped sideways. It was from Turn 4, down past the flag stand before I ever got it back straight. I was all over the place. I came back in and said, "What in the hell did you do?"

Harry said, "I took the damn spoiler off. Now that's loose."

He made his point.

Harry was great. Him and Tim Richmond? That was the ultimate combination. Tim Richmond was the ultimate when it came to car control, and Harry would give you just as much of that as you could stand. They were perfect for each other.

*In victory lane after the 1973 World 600 at Charlotte
with car owner Nord Krauskopf.*

16

Back Off, Baker!

As NASCAR and Chrysler locked horns over rules and engines and just about everything else in 1974, Buddy found himself virtually on the sidelines until he got a phone call from Bud Moore, the colorful owner of a Ford team. Moore was a decorated World War II hero who took part in the D-Day landing on Utah Beach at Normandy. He didn't lose too many arguments, either.

Chrysler wanted me to sit out until the rule situation got better.

Bud Moore called me and asked me if I wanted to drive for him. Chrysler wanted me to go run the Pike's Peak Hill Climb and stuff like that for them.

Bud said, "You're a race car driver. What are you messing with stuff like that for? That's not you. You want to go to Charlotte?"

I said, "Yeah, but I have a contract."

Bud said, "Well, they're not going to race this year. Do you want to sit around and do nothing?"

I told Bud I wouldn't drive for him if I was going to take somebody else's spot. I had already been through that. Bud was pretty blunt about things. He said, "My driver's gone, whether you come or not."

There was a driver who drove for him once. Bud was a father figure kind of guy. When he told you something, it was going to wind up his way. You could argue all you wanted to.

This driver wanted to change the car over to short trailing arms. They hadn't had a finish worse than third in the three races the guy had driven for him.

Bud walked into the shop, and they were standing around in a group in the back of the shop. He went over and said, "What are you guys doing?"

They said, "We're thinking about changing this thing over and putting short trailing arms on it."

He said, "I'll tell you what, you pick your best car out of here and take it home and put short trailing arms on it. My car's going to have long trailing arms on it."

And the argument was over.

He got on me at Ontario, and we'd won the race. It was the only time I ever won a race for anybody who wouldn't speak to me afterward.

It was at Ontario in 1975. I was out there and David Pearson was running second. We were like 28 seconds ahead, and Bud told me to back off and take care of the car.

"There ain't no need in you running as hard as you're running," he said.

*Sometimes I tried to argue with Bud Moore when he
was my car owner, but I rarely got anywhere with him.*

So I backed off. But all that did was give me a great run into the corner. I could get back into the throttle earlier. My lap times started dropping, and I was actually going faster.

Bud comes on the radio and says "You hard-headed S.O.B." and he started calling me some names I won't go into.

I said, "I swear I backed off! I don't know what's happening."

He said, "Well, back off more."

I did. I started backing off earlier, and it freed the car up even more and turned it into a rocket. I was flying down the straightaways. And the lap times went down some more.

Bud just took off his headset and flung it out into the infield.

"Somebody else talk to that crazy sum-bitch," he said, "he won't listen to me."

I won the race, and they gave us a new car for winning, as a bonus. I thought if I said I was going to give that car to his son he would get over it. He wouldn't come to the winner's circle or speak to me.

I went down to the shop in Spartanburg, S.C., the next day after we got back. He had his door shut. That wasn't a good sign. If you raced for somebody and you went into the shop and he knew you were there and still kept his door shut, it wasn't good.

I finally opened the door, and he was looking out the window.

"How are you doing, Bud?" I said.

Not a word. Silence.

Then, he slowly turned his chair around.

"Let me tell you something," he said. "You have got to be..."

I stopped him.

"Before you say anything," I said, "I kept backing off and kept backing off and the car got faster. It got a better roll through the center part of the corner, and with the gear and all we had it was perfect. I let off and you wouldn't believe it."

Bud just looked at me.

"You can talk until you fall dead right on that spot, and I am never going to believe you," he said. "You're a liar."

Finally, about a week later, he started talking to me again. He was mad; he just absolutely thought I just wasn't going to listen.

That's not what he was about; he was the boss. That was his race team, and if you started not doing what he told you, you were in trouble.

Bud and I went through some arguments; we were both headstrong.

I had won three straight races at Talladega, and we had a fourth one won, too. I told Bud the car was strong enough that I could come in and get gas and be back in the lead in five laps.

"Aw, you've got plenty of fuel; shut up," Bud said.

I ran out on the white-flag lap.

17

The Silver Fox

David Pearson and Buddy Baker had similar careers in Winston Cup racing in that both spent much of their time driving for teams that didn't compete in every race.

Their teams would enter the major big-money races and try to win them, leaving dozens of midweek short-track races to others. Baker spent much of that part of his career augmenting his income testing tires for Firestone, learning his way around the sport's faster tracks in the process.

Pearson did all right for himself, though, winning 105 races to rank second all-time to Richard Petty. Buddy, however, contends that all of the bad luck he had in fast race cars evened out in all of the good luck that Pearson had in his career.

I love David Pearson. Every time I see him I call him the luckiest man alive.

He'll say, "What do you mean, lucky? You make your own luck. You have to be in the right spot when things happen."

I'll say, "Don't give me that crap."

I started on the pole at Talladega once in the Nord Krauskopf-Harry Hyde car, and Pearson was on the outside pole. We started off, and his car wouldn't run a lick. He'd fouled the plugs out.

We made three laps, and he was just before getting lapped but was still a little bit ahead of the mess when a couple of lapped cars got together on the backstretch. It ended up being a 28-car pileup.

I knocked the motor slap out of my car, and Cale Yarborough went over me in the air, still accelerating. Bobby Allison hit James Hylton so hard it cut his car in half. I couldn't begin to tell you all of the stuff that was going on in that wreck. One car had been in the pits and came backing down through there and almost hit Cale and me running backward at 200 mph.

After all of that, I looked up and here came Pearson running through there at about half speed, puttering along. He drove through it, went in the pits and changed two spark plugs.

He went around three more times and came in and changed out all of his spark plugs under yellow. And then he won the race. You tell me who's lucky!

One time at Michigan I had a problem in the pits, and he was about 15 or 18 seconds ahead with just a few laps to go. I was catching him like four-tenths a lap. I pulled up on his back bumper, and NASCAR decided it was too dark and threw the checkered flag. And Pearson is telling me he's not the luckiest man alive?

But there's nobody I have more respect for. You talk about a driver who knew how to win—he did. He would run just as hard as he needed to until it came time, and then there was nobody any better.

We were sitting around a table somewhere one day, and I was about to be inducted into a hall of fame or something.

He was sitting there and he said, "Well, by God, you ought to be."

When a competitor like David Pearson says something like that about you, that's pretty nice.

18

A Ghost of a Chance

By the end of the 1977 Winston Cup season, Buddy
had won 13 races for five different car owners. But he didn't
win that season in Bud Moore's cars, and he decided it was
time for him to look elsewhere if he wanted to accomplish
the one goal that had begun to haunt him.

Buddy had won both races at Talladega in 1975 and
followed that up with a victory in the spring race there in
1976. That proved he knew how to handle a race car at the
sport's fastest tracks. But he had never won at Daytona, de-
spite having had cars that clearly seemed capable to doing
just that. Particularly frustrating had been his misfortunes
in the Daytona 500, the sport's biggest race and one Buddy
wanted desperately to win.

The joke was that Baker had won the Daytona 475
several times but never had made it to winner's circle at the
end of that 500-mile race. It wasn't exactly funny to Buddy,
though, and when he left Moore's team after 1978 it was

because he felt he could best chase his dream of getting a victory in that race for another team.

Right from the start of my career, I took rides that a lot of people would say, "Have you lost your mind?" They wondered why I would leave one team to go to another. I always felt there came a time when you'd gotten what you were going to get out of a ride.

It was a challenge to me to go to somebody and have them tell me, "You know, we just haven't had any good results. What do you think you can do with our team?" I loved helping them become a front-runner. That was as much fun as taking the best car and winning with it.

Leading the way in the "Gray Ghost" in the 1980 Daytona 500. It should have been the second straight year I won that race, but an ignition problem took us out in 1979.

My philosophy was that they put a number on it and it didn't have a phone number on the door or a meter in it. It wasn't a taxi. They built those cars to race, so I did.

I quit Bud Moore because I felt like we'd done what we were going to do together. We had a race where he had lapped the field at Daytona and the little ball on the end of the accelerator had pulled out. The cable got pulled up through there; I could only run about half-throttle and still had a dead heat with Cale Yarborough for third.

We started having this and that happen, but I would be lying if I said anything other than I was obsessed with winning the Daytona 500 to the point that it was agonizing. That became something that I had to do.

With Harry Hyde I once had a 27-second lead with just a few laps to go and had a right-front tire go flat on me. I had to sit there on pit road and watch a guy come all the way around the front straightaway and pass me with just a couple of laps to go. The tire just ripped the car apart when it came loose.

I left Bud Moore mainly because I thought I had a better chance of winning the Daytona 500 in M. C. Anderson's car. We didn't even have a garage; we built the cars in an old filling station with two bays. But we had people who believed in each other.

I was leading the race, and the Daytona 500 victory was just a few laps away. I was already cutting up with the crew, telling them "I can't believe my first Daytona 500 win is going to be this easy."

Then the engine let go on the front straightaway. The motor had run out of oil, and I had to sit there on pit road and watch Bobby Allison in the car I'd just gotten out of—Bud Moore's car—come from a lap down and win the race.

I wanted to jump off a bridge.

It seemed like I could fall out of bed and win at Talladega, Daytona's sister track. I actually was better at Daytona than I was at Talladega.

I was going to do whatever it took, and I thought I could win with M. C. Anderson. It was a situation where they needed a driver and I needed to make a change. And I did, and I came within about four laps of winning the 500 my first time out.

Harry Ranier wanted me to drive for him the next year. He sent somebody over to talk to me, and I said I wasn't interested. I was driving for a guy I absolutely loved. M. C. Anderson was a very nice man. We hadn't won, but I knew the possibilities were there.

But Harry offered me a deal I couldn't turn down— a chance to drive the 28 car. M. C. then came back and offered me something that I just couldn't believe. It was way more than I was making.

I told him, "Nobody deserves that; I'm not worth that. Besides, I've already shook hands with Harry Ranier that I would drive the 28 car next year. And that's as good to me as a contract, "

I might have won the Daytona 500 anyhow, but the first time I drove that 28 car I thought, "Good God! Where did this come from? You mean I have been out here racing against this?"

We went to Daytona and won the first Busch Clash ever in 1979. We won the pole for the 500 and we won our 125. We absolutely had the best car down there. NASCAR was all over it looking for something.

Race morning, the crew found a strut to the back bumper loose, and they welded on it. It knocked the ignition out, and we didn't know it. They dropped the green flag and the car never came up to speed.

I came down pit road and I told them, "It must have broke a timing chain or something; I don't know what's wrong."

They jumped in and unplugged one ignition system and unplugged the second system, and a guy got excited and stuck it right back where it was. It wouldn't crank.

We got home, and Waddell Wilson, who was the best engine builder there was, said, "I just cannot believe that motor went."

He reached in there, unplugged the first ignition and plugged the other one in, and the motor cranked. In trying to hurry to get me back out, the guy who'd reached in during the race had unplugged the bad one and then plugged it right back in instead of switching it.

I shouldn't have been the one trying to analyze the problem. I could have made up all kind of laps that day with no problem. In the 125-mile qualifying race, Cale was leading after the pit stop. I came in and went out too slow, and when I came out I was 11 seconds back.

When I ran him down and got behind him, Junior Johnson says he told Cale, "Baker's coming, Baker's coming. Make sure you get in the draft."

Cale called him back and said, "Junior, he's gone."

That car was a rocket. I knew that if I stayed with that team I'd win the Daytona 500.

We kind of cooled it the next year. We came back and won the pole, and NASCAR was all over that car. They were checking everything.

It was the car everybody called "The Gray Ghost." The first year I drove it, in practice, two or three people said it nearly scared them to death, because they would look back and the car was gray on top—it blended in with the pavement. I was so quick that when we got on the straightaway I got on them so quick it was like I came out of nowhere.

D. K. Ulrich and a group of them got together and told NASCAR, "You've got to do something, because he's scaring the living hell out of us." They made us put Day-Glo strips on the front of the car so people could see it. That's why we called it The Ghost.

The night before that race, I couldn't sleep. I knew what I had. I had sandbagged all week so we wouldn't get grief from NASCAR and the other teams. The car was so quick it should have been looked at. I had saved that car all week, because if I had put a dent on it I think I would have cried. I was so proud of that car.

The night before, I was in bed thinking about the start and who I wanted to run with and stay away from. I didn't get a lot of sleep that night—I'll bet I didn't sleep three hours total. I would wake up and think, "I need some rest."

About 5:30 I got up and took a little walk. I came back in, sat down in the chair and went right to sleep.

When I went to the race track there was this guy on the gate, a guard, and for some reason they have a knack at Daytona for finding the most ornery people to put on the gates. You can't talk to them. You can't tell them anything. Nothing matters. I think they go into neighborhoods and ask people who the most ornery person is in the whole neighborhood and go hire him. I've had passes that were supposed to be good anywhere down there and the guard would say, "Anywhere but here."

I would be in my uniform to go in the garage area, and they would stop me and ask for my pass. That griped me. That morning, this guy stopped me and would not listen. Finally, I told him, "Hey, I've got briefs on, too."

It made me mad, but it got my mind off the race just for a minute, and I think that helped.

Nobody but me knows how fast that car really was. We ran third in the 125 or something. But when they dropped the green flag in the 500, I said, "There ain't a thing they can do about it now; here we go!"

It was a no-brainer. All I had to do was to keep from making a mistake. It never was really a contest. I had won the Daytona 500.

I couldn't sleep a wink the night after the race. You would think after 500 miles at Daytona you would need some rest. I went to a couple of things after the race that night, then went back to the hotel and got in the bed.

I would shut my eyes and they'd pop right back open. "Daytona 500! Golly, can you believe it?!?"

I twisted and turned and I finally said, "This is stupid. I am going home."

I got up and got in the car to drive home by myself. I was riding along with the radio on. It was the middle of the night, and just as I crossed over the Georgia state line I came over a rise in the road and my radar detector started hollering at me.

I said, "Oh, no!" There he was—the state patrolman—sitting in the edge of the woods.

It must have been 2 o'clock. I didn't even wait for him to pull out. I just went ahead and pulled over, and he came on up behind me and stopped.

This state trooper walked up and saw me and said, "Well, Buddy Baker, I'm going to tell you something. You are my favorite driver. I always pull for you. But you have some of the most rotten luck, and here's another example of it."

I said, "You're not going to give me a ticket, are you?"

He said, "Absolutely. You do your job pretty well, and I do mine."

He did give me a break on the speed. I have no idea how fast I was going.

Sometimes you live long enough to have your dreams fulfilled. After winning that race, everything else was a bonus plan.

But if I had never won the Daytona 500?

I'd be a raving maniac today, or even more of one than I am now.

19

Hit and Run—Part II

A few more notes and quotes from Buddy about his career and some of the people he raced for and against.

I had 43 second-place finishes in my career. Look at the top fives. If I had as many top fives now, I'd be riding around in my private jet.

I got beat sometimes, sure. But there was a time in the sport when attrition played a big role. Some people just rode around and wound up being opportunistic, and things happened that suited their driving style.

Cale Yarborough drove for Junior Johnson at the right time. So did Darrell Waltrip. I was offered an opportunity to drive for Junior Johnson once, back in 1968, and I have wondered if I had done that how many more races I might have won. His cars were more durable. Junior was a master of set-ups.

I was pretty good on the short tracks. I won a race at Martinsville in 1979 with Harry Ranier. The team looked at my driving style and thought and thought about it and decided there was only one way to do it. So we put the Daytona motor in it.

When I would mat it in the center part of the car it wouldn't break the wheels loose because it was a top-end motor. But when I got to about the flag stand on the straightaway it was like somebody had hit you at a stoplight. It would just take off.

The speedway in Ontario, Calif., was almost a carbon copy of Indianapolis, and they had to dig it up to keep me from winning a lot there. Every time I went there there we ran up front all day long.

We went there one time with Bud Moore without a sponsor, and this guy came up in a pair of old blue jeans and an old Madras shirt.

We were on the pole, and he walked up and said, "I'd like to sponsor you all." He said he was with Norris Industries, and we thought that was something like Honest Charlie's or something.

He asked what it would cost, and I went over and asked Bud.

He said, "Tell him $10,000 for tomorrow."

He said, "Okay." Just like that. He wrote a check, and Bud told me to call the bank and see if it was okay. The way the bank answered me was funny. They said, "Yeah, he can cover it."

He said, "If you win the race, I will just sponsor you next year." And we won the race. I stayed over to see what Norris Industries was, and did I have a story to tell Bud when I got home.

I was riding along in a cab, and I said to the driver, "Man, look at these buildings. What is all of this?"

He said, "Norris Industries. They make all the missiles for the government and all of this."

I said, "Holy cow!"

We were going to send this guy for hamburgers for lunch at the track. I walked into his office, and it must have been 300 yards long and 300 yards wide. He had a guard in there with him.

He asked me if I had a figure for the next year, and I told him what it was. It was done, right then. There wasn't any negotiating about it. He ended up sponsoring Bud for two or three years.

That taught me that you should never judge somebody by the clothes they have on or the way they look.

It's a shame that race fans today don't have any idea how good some of the guys who drove and won races back in the 1950s and '60s were.

Rex White is a perfect example. Nobody knows him; he can go to the race track now and get lost in a crowd of two people. At one time, he was the champion and was nearly unbeatable. He was way better than people will ever know. Today, if you went up to the average race fan and say Rex White's name, they'd go, "Who?"

Nelson Stacy was a great race car driver. Fred Lorenzen was Jeff Gordon. A lot of people didn't like him because all of the girls did. He was Ford's No. 1 driver, even though Pearson was on the same deal.

Lorenzen was the golden boy of that era. People don't know how good he was. He made so much money so quick, he was in and out. Then he tried to come back, and you can't do that. In his time, he was good. Had he stayed in the sport, his numbers would have been way up there.

I never raced for a salary. I wanted to earn my money. I'd get money up front to run for the year. I didn't want a salary; I didn't want to owe the company. It was like I was on commission.

We didn't know who Ken Norris (back row, to the right of me) was when he offered to sponsor Bud Moore's team in the 1975 race at Ontario, California. The next day, I found out Norris Industries was a big deal.

Once I started winning, I never drove a race car for less than half of the purse. I went to a race team one time, and they told me what a great team they were and how they could get anybody for 35 percent.

I said, "The people you can get for that know what they're worth. I'm going to get half."

They called the next day and I had the job. They had to try, I guess.

Darrell Waltrip came into NASCAR telling everybody he planned to take Richard Petty's place. He might not have done quite that well, but he wasn't half bad.

The toughest driver I ever drove against? Harry Gant would have to be right at the top of that list.

He wasn't mean or anything, but I will bet you if you ever cranked up that old country butt he'd be something to deal with. I watched him take a ball-peen hammer with the long handle on it, take his finger and his thumb only, and bring it up to touch his nose, then put it back down. Other people couldn't even budge it.

When he's doing carpentry work, they say he can walk the long beam the length of the house on his fingertips. He's a stout son of a gun.

I never saw Harry mad but one time, and luckily the driver understood Harry was coming after him and took off. He locked himself up in the front part of his truck.

Cale Yarborough was that type of person, too, on the race track. Off the race track, Tiny Lund absolutely had to be the toughest. He and Larry Frank. Larry was an all-services boxing champion. He could stick with Tiny, but as far as physically strong they were the strongest I ever raced against.

Darrell Waltrip once was the most unpopular race car driver in this part of the world; there's no question about that. When he would drive to the race track, people started booing him when he turned off the bypass.

The first time I remember him, we were at this place called the Castaway in Daytona. They had a radio show, and none of us knew him from beans. We didn't have a clue who he was or what he represented or anything else.

They introduced him as having been quite a winner at Nashville, and he was driving the Terminal Transport car, number 17, and this and that.

Darrell came to the microphone and said, "Hello, everyone. I'm Darrell Waltrip. I'm the guy who's taking Richard Petty's place." Just like that.

People said, "Are you kidding me? Shut up!"

He came on the scene, and he was like Muhammad Ali. He'd brag, but he backed it up. People absolutely could not stand him.

Then Rusty Wallace turned him around in The Winston at Charlotte and it went 180 degrees. He went from the most disliked to one of the most popular guys out there. All of a sudden, he was the guy.

Cale had given him the nickname "Jaws." Darrell and Cale had some classic stuff. He went after Cale one time and spun himself out. He sat right in the infield at Michigan and blew his motor up trying to get out of the mud to go after Cale.

He knew how to work to get recognition, but the way he did it kind of ticked people off until you got to know him.

Once you got to know him, though, that was just Darrell.

———————

Several years back, Rusty Wallace asked me to go to Daytona and work on a project in testing with him.

He ran the morning session, put me in it, and told me to tell him what I thought about it. I took it out and came back in. He was looking at the times and said, "I'll work on it and you drive it."

He worked on it all afternoon with me driving. I wanted to see if he was paying attention, watching what was going on, and in Turn 1, instead of running where I usually run, I just went up the race track and went back down.

I came in, and we were going through the computer data and everything. He goes, "Right here in the middle of the corner the left-front looks like it was kind of coming up, and what were you doing messing around down there?" I was just checking him out.

I've heard many people say things about Rusty, but you never know how kind he really is about everything.

He comes off pretty fiery, but when I work with him I get nothing but pure respect. He has just been super with me. When I work with him I feel like I am a part of the team, and there have been times where people didn't make me feel that way.

When I owned my own race team one time, I had a driver, and I told him, "I'm going to have to let you go. I hate to do it. It's nothing to do with your driving skill because you've got it all, but if God was to kick a rock out of heaven it would fall to earth and hit you right square in the top of the head. I've never seen anybody with as bad a luck."

Rusty's had good luck, too, but he's run a lot of times where he's been good enough to win and hasn't.

I brought Jimmy Spencer into Winston Cup. He drove for me when I started my own team. He's a hard-headed son of a gun, and I thought I was hard-headed.

After I got hurt, I was kind of confused about what to do. I didn't know whether to keep the team or not.

Spencer came by and said, "Why don't you put me in that car?"

I laughed. He and Dale Jarrett were fighting about every week in Busch, and I said, "No way. I don't need that kind of trouble."

But I got home and started thinking about it and figured he was hustling that Busch car pretty good. I called him and told him to come over and talk to me. He came over, and I told him he had the ride. Then I met the rest of the family. He could do no wrong in their eyes, and his family is really close.

Jimmy is a delight to be around, but he's another guy who hasn't had a lot of racing luck.

We went to Daytona one time and we qualified 15th or something like that.

"It won't keep up; it won't run," he said.

I said, "Jimmy, drive this race car and don't worry about your practice times. People who win practice don't run well."

He went home Saturday night mad. "The car won't run."

I told him that after five laps he was as good as anybody, after 10 laps he was better than most, and after 15 laps he was absolutely perfect.

The next day, I told him before the race the car would be good in traffic. The race started, and he was picking them off.

He got up to second, and I said, "Jimmy, how many cars are in front of that no-running son of a gun?"

Benny Parsons is one of the best mechanics there ever has been in this sport. He was down in Ellerbe, N.C., working a car down there with Travis Carter and Jake Elder and a few others. Benny was in there building the gears and working on the front end of the race car.

To win a championship the way he did was remarkable. He did what it took.

When he went to work for M. C. Anderson, he told him, "If you're looking for another Buddy Baker, you're not getting him. But I will win you some races when it's right."

When Benny had the best car, he won the race. A lot of people couldn't do that. With Benny, one of the best things you could say is if you had a third-place car, you got third. If you had the best car, you won. He knew his abilities and didn't overstep them.

I don't know of one person who ever had a run-in with him, not one person who ever got out of the race car and said, "I am going to go after Benny Parsons because he did me wrong."

For most of the teams I drove with, the championship wasn't that big of a deal. It was a part of it, of course, but most of the teams I ran for ran limited schedules. I ran every race for Bud Moore one year and finished fifth in the points, in 1977.

I never put a concentrated effort toward winning a championship. To some people that was everything. I was competitive, and I had a life, too. I got to do things some people didn't have a chance to.

They call the championship the "big picture." In the past 10 years, that big picture has gotten a lot bigger. They didn't have the $2 million payouts back then they have now. Today, it's a treasure beyond a treasure, but that came from Winston's involvement in the sport.

I think everybody would have loved to have had a championship. But there are a lot of great race car drivers who never won a championship. They were still great.

I've always looked at it this way. I was dealt a pretty good hand in my career, and I can't miss something I never had.

20

The Man in Black

Buddy and Dale Earnhardt were practically neighbors in North Carolina. Their fathers had raced against each other for years on the dirt tracks around the Carolinas. Earnhardt and Buddy shared another level of kinship, too; their ability to understand what it took to race at Daytona and Talladega, the two NASCAR tracks where both drivers learned their lessons about the draft quite well.

The first time I met Dale Earnhardt in my life, he came up to me, and he was a little, thin thing. It was over at Robert Gee's shop next to the speedway in Charlotte.

He came up to me and said, "Buddy Baker, I'm Dale Earnhardt."

I said, "Well, I know your dad."

I didn't know who he was. There wasn't any reason for me to know him. He was still running on short tracks.

He said, "I've got tell you something. I've watched you run, and if we ever get down to the last 10 laps, I'm putting you out."

I said, "Thanks for the warning; I'll be looking for you." We all laughed about it. But the first time we got down to it, I remembered that.

In 1980 at Talladega, I came out of the pits 27 seconds behind Earnhardt. They started giving me lap times, and I started chipping away at the lead.

They said, "We've got second sewed up; take it easy."

I said, "I believe I'm going to keep running it wide open, and if we don't win we don't win."

You could hear the crowd getting louder, and louder and they kept giving me the times. After a while, I said, "You can be quiet now. I see him."

I caught him with two laps to go. I knew there was no way in the world I wasn't going to go by him right then. If I had waited until the white flag lap, I wouldn't have gotten by him. I passed him, and he thought he was going to rip right back by, but I understood the draft just about as well as anybody.

I started going up the track, and he stayed right on the bottom. I beat him by two or three car lengths.

After the race, he said, "I can't believe you gave me all of that room down there."

About four months later, Dale came up to me and said, "You sorry dog. I beat a guy doing the same thing you did to me. All of that room you gave me wasn't doing anything but putting a wall of air up in front of me, and I ran right into it."

I just said, "Well..."

He was a little on the devilish side. He'd look at you and grin and start inching over. At Daytona, with the front-end settings you have, the last thing you want to do is put the right-front against the flat wall. It will climb it.

Just off the dogleg there on the front straightaway, he started inching over. I said to myself, "Does he really believe I am going to let this happen?" He kept coming, and I just turned left.

When he finally straightened out, he realized he would have to find another way to put me in the wall.

A lot of people have heard all the hard stories about Earnhardt. But he'd come over here and we'd get in his bass boat and ride down the lake. He would sit there and listen for hours about what you wanted to do on major speedways and in the draft. He'd do it with Bobby Allison, too. We'd talk about stuff like that. He paid attention.

One time he told Bobby Allison, "You taught me everything I know."

And Bobby said, "Yeah, but I didn't teach you everything I know."

It was obvious to anybody who watched him run that he was pretty soon going to be one of the most famous of all of us. But when I first met him, he would have had to borrow 75 cents to have a dollar.

There were times when he had enough money to go one way to the race track. Ralph was a tough guy like my dad. He would give Dale enough money to go one way, and Dale knew he had to finish the race to have enough to get home.

I am not sure that wasn't good. It gives you a background in what everything's about, and you appreciate the good things more when they come. Dale Earnhardt borrowed parts and scrounged to get started, but once he had everything, if he had something you needed, it was yours.

Up to the very end, I think he still knew what a dollar was all about. When you start with nothing, if you start dealing with big money, you're wary. If you have everything to start with, sometimes you don't appreciate it. I will guarantee you in Earnhardt's mind he always remembered those days when he had to win the money he needed to come home.

Two days before he got killed in the 2001 Daytona 500, he raced in the International Race of Champions and got into it with Eddie Cheever Jr. Earnhardt got off in the grass, went 200 yards, came back up on the race track and barely lost a spot. He never lifted. He did things an average man couldn't do.

When I saw his wreck in the 500, I realized it was bad. But I fully expected him to climb out of that race car. I was watching it on TV at home. It was the first year after CBS had lost the rights, and I wasn't down there doing the race. As much I loved doing that race on TV, that was one time I was thankful I wasn't there.

Even after you heard the news, you were thinking, "There's no way. Somebody's got to have some bad information."

I was down there doing the race for CBS in 1998 when he won the Daytona 500 and people from every team on pit road came out to congratulate him. It took me 18 years to win the 500, and it was his 20th. I knew what it meant to me, and it was good to see the people who maybe two months before that were mad because he'd run into them and cost them a spot, stand out there and do that. They realized he deserved to win it, and they felt like they were part of it. It was like John Wayne winning another battle.

I don't cry much, but that would almost choke you up. I'd heard some of the real know-it-alls say he'd never win it. And he did. Having won it, I appreciated what it meant, and I really appreciated somebody who won it seven times, like Richard Petty did. They would have to tie me down if I'd won that race seven times.

21

Greener Pastures

As good as things were for Buddy with Harry Ranier's team, racing has a way of making good things come to an end. By the middle of the 1980s, Buddy would start his own race team along with Danny Schiff. "I decided I wanted to have a little bit of say in what happened to me," Buddy said. The road toward that decision, however, was sometimes a bumpy one.

I left Harry Ranier because some things started happening to us that shouldn't have been happening. Our chemistry had been so good there. Our crew chief lost his job, and that sort of demoralized the team. Once that starts, the team isn't going to be as stable.

When personality conflicts like that start, it's hard to stop them. I am sure there were some people there who didn't mind seeing me go, either.

I was told one time I would be in that car until I turned cold. Then, it started to snow in July.

Harry Ranier and I have always been great friends, but I didn't agree with some of the decisions he was making. Once you have that feeling it's best to go ahead make a move.

Near the end of the season one year I had won a race at Talladega and was walking across the infield, and somebody asked me, "What's this I hear about Bobby Allison driving your car next year?"

I laughed it off. I'd just won the race.

I went over to the truck to change clothes and saw Harry. I said, "Guess what? I just heard the damndest thing you've ever heard. Somebody said Bobby Allison might be in this car next year."

We used to walk right out onto the track and sign autographs for the fans—but there weren't as many of them to sign for as there are today.

Harry said, "Well, we have been talking to him."

The light went on. We won six poles and two races in 1980, and the next year I was gone.

I was offered the opportunity to go to Hoss Ellington. Hoss was in need of a driver, and I needed some place to go. It wasn't something that was ordained or really something that should have happened. It's hard to be satisfied with a new ride when you know you've left the best race car out there, and a lot of people don't want the obligation of trying to live up to what you just left.

If you wanted to laugh a lot and enjoy yourself, Hoss's team was the place to be. That was a great bunch of guys working on that car. They had a great time, and so did I. They had Donnie Allison before me, and I had been with Ranier, and it was sort of a transition for both of us. I had good race cars when I went to the track and we ran well, but we kept having little things happen to us.

I was so far ahead at Pocono one day that it started raining in Turn 3 and the second-place car had enough time to slow down and make it through there. I got there first, and it was raining. I said, "Somebody over here has a grill going, because I can see the smoke." That's how hard it was raining. I went in there wide open and ate that wall up.

I was so far in front that I got to the finish line where they red-flagged the race and still had the lead. I said over the radio, "Hoss, all I can tell you, pal, is we've either won this race or we're done for the day if they restart the race."

About the time I said that, the sun popped out and they started drying the race track. When they cranked

Leading the way in the 1983 Firecracker 400 at Daytona in the Wood Brothers' car. This was my last major race victory.

the engines up, they took my car to the garage area on a wrecker.

When I left Hoss Ellington in 1982, I went back to Harry Ranier's team, and I learned that's one thing you can't do.

The next year, I went to the Wood brothers. Valvoline had been the sponsor when I was Hoss Ellington's, and they were going to make a change. They went to the Wood Brothers, and I'd been associated with Valvoline, so I wound up there.

That was a fun time, too. They told me a couple of years later after I was out of the ride and had started my own team that they wished they'd had their cars as good as they did then when I was in it.

We were so far ahead at Talladega one time that I turned the car off in the backstretch to coast in to the pits and stop because we had no brakes. When I stopped, they serviced the car like nothing was wrong and I told them the brakes were gone. I went back out and was still in the lead with no brakes. Can you imagine having a lead like that at Talladega now?

I won my last major race for them at Daytona in 1983.

If I had stayed with Harry Ranier, I probably could have won another 10 races. But when things start to go away, I would rather have been happy than rack up big numbers. I still wouldn't go back and do things a different way. Rather than have friendships turn bad, I would rather go somewhere else. But I do think I could have worked a little harder at it sometimes and made some things work.

22

Big Bill

NASCAR founder William H. G. France helped guide the sport through its early years and lay the foundation that his sons and grandchildren are building on today. "Big Bill" France could be a benevolent despot, but he was a despot nonetheless. He ran NASCAR by the sheer force of his will at times, and it wasn't often that he alone wasn't enough to carry the day.

He was a tough son of a gun, and there wasn't any winning of the argument with him. At the same time, not many people knew the real Bill France.

I was sitting in a restaurant in Daytona one time having dinner, and Bill was over at the next table. I don't think he ever carried any money on him.

When dinner was over, I saw him fumbling around, and then he started looking around a little bit. He came over to me and said "Have you got any cash on you?"

Now this was a time when "a little cash" was exactly what I had. I did have enough to cover it, but I knew that as busy as he was he would never remember that.

The next day, I was up under the race car working on something, changing the gear, and his limousine pulled up in back of my garage space.

"Where's Buddy?" he said.

I wheeled out from under the car and he said, "Here's your money."

Everybody there was wondering what he was giving me money for. I just laughed and said, "Yeah, I've got him right where I want him now."

He brought people in from all over racing all over the world, and he'd get them in the Daytona 500. He understood that having a little international flavor to it was good. He was a tough old dude who knew that to make the sport grow he had to do some special things.

One time he called me and said, "We want a Plymouth in the race down here; would you build one?" I said I would, and he told me to go buy one and get it ready and they would make sure everything got taken care of.

We went down and ran second to A. J. Foyt in the 500. That car had a windshield in it you would not believe. A city bus doesn't have a windshield that big in it. We were way back, but we came up to finish second, and he had a Plymouth in the race when Chrysler was boycotting. That was 1965.

One time at Talladega he came by, picked up this one driver, took him over to a clothing store, and bought him some suits and a bunch of other stuff. Bill realized guys like that were valuable to the sport and that they did the best they could with the money they had. He just wanted to do it. Nobody asked him to.

On the other hand, I was there when the Professional Drivers' Association—the driver's "union"—boycotted the first race at Talladega in 1969. It was new, and we got down there and there were some real problems. You could go out there, run six laps, and shred the tires so bad it looked like spaghetti hanging off. The cords would be hanging out if you ran hard for six full laps.

Celebrating a victory at Petty Enterprises with my son Bryan and Maurice Petty. Notice the PDA patch on my uniform. That's the Professional Drivers Association, a driver's "union" that brought about a boycott of the first Talladega race because we had concerns about the safety of the tires on that new, high-speed track.

Up to that point, the PDA hadn't really tried anything. We stood up and said we needed this and that fixed and that the tires were just not safe. We also said it would be better just not to have the race.

Well, Bill France came down there, crawled in somebody's car, ran around the track at 140 mph and got out. "See," he said, "you can run here."

We were sitting there with 200 mph cars, though, and we all agreed it would not be safe to run. There were some drivers, like Bobby Isaac, who said they were going to run, no matter what, and they shouldn't feel bad about it. But I had given my word that I would do what the group did.

I was racing for Dodge, and I called my bosses and told them we had a situation down there that could turn nasty. I knew that if I was out there trying to run 150 mph that Cale was going to come by me running 155 and I would step it up. Richard Petty would then start chasing us, and we had cars that would run 200 mph. We all agreed we weren't going to run.

Once they started arguing, I just kind of stepped back and watched and made my own decision that I'd given my word and was going to stand by it. It might have been the best thing that ever happened to Talladega Superspeedway. If we had run, it would have just been another place down here in the Southeast. But because of the controversy and because it was the first time the drivers stood up together, it got all kinds of national attention.

We went back the next year, though, and had one of the best races you ever saw at 200 mph. We made the right decision to leave, and it was the right decision to go ahead and have the race.

The only bad feelings that came out of that were toward people who said they would go with us and then changed their minds on the day of the race and ran. Later on, I remember one driver who did that who came to a meeting we had a couple of weeks later saying, "If I ever drive for you, I wouldn't do you that way."

Lee Petty said, "Let me clear the air, young man. I will promise you that you will never drive for us."

I told everybody after the PDA did fall apart that, from then on, when they drop the green flag I am going to be there. I had lived up to my end of the bargain, but from that point on, I was racing.

To get back to Bill France, though, during that weekend LeeRoy Yarbrough complained that he was getting vertigo when he raced at Talladega. Four years later, I was standing in the garage at Talladega and Bill France walked up to me.

"How's that vertigo?" he said.

23

The Not-So-Gentle Giant

One of the nicknames that stuck with Buddy during his career was "The Gentle Giant." Like many of the nicknames coined by sportswriters of the 1950s and '60s, Buddy says that one was a stretch.

I don't want people to get the wrong impression of me, but this nickname "The Gentle Giant?" Where in the heck did that come from?

They called me "Bouncing Buddy Baker" one time, from an announcer down at Darlington who just started calling me that.

I asked somebody one day, "Do I bounce?" They said no. I never did understand that.

But the Gentle Giant? I wasn't Gentle. Ask my wife. She hears that and just laughs.

My dad came from an era where there was absolutely no tolerance for anything. If you bumped somebody you were going to get the devil knocked out of you. Period. If you went down to argue about it, you had better have your fighting face on, too.

In my era, we respected each other, but there were times when things got carried away. I've been mad enough that going into the last lap I had no intention of making the last corner. If I missed the guy I was going after, I was not going to make it. I was usually lucky enough not to miss.

One thing I would never do, though, is try to hurt somebody. I had a bad temper, but I wasn't the only one. We weren't afraid to express how we felt. That's not a bad thing, I don't think. If you're mad at somebody, it's better to get it out and deal with it instead of waiting six months and catching the guy unawares and maybe hurting him.

If they hit me and they saw me coming, they knew they weren't going to have to wait until the next week.

There was this one guy I had some trouble with. He was supposed to be the kingpin at this one track, and I was traveling all over the country to race.

Sometimes the local hotshot didn't like you coming into his own backyard and racing him. That night, it got down toward the end, and we made contact with about three laps to go. He went out and I didn't. The people in the grandstands didn't like that at all. I couldn't think of anything to do, because I certainly couldn't get out of the car and fight them all. I just put the car in low gear and went around and around, flinging dirt on them until there wasn't anybody around me any more.

Another time I had to have a police escort to get out of this track. We were going through there and one guy called me a name. I turned around and called him a name. The local policeman looked at me and said, "If you don't shut up, I am going to shoot you myself."

I am not proud of times when I acted like I didn't have any sense. There were times I just turned the wrong way on the track and went after somebody, and as soon as I did it I knew it was stupid. Your adrenaline is maxed out and somebody hits you or takes away a chance for you to win the race? It's hard to just go, "Heck fire darn, that's too bad."

Jeff Gordon showed that in the final race of 2001, when he went after Robby Gordon, he showed the world what it's like. Who would have thought Jeff Gordon would just go plow into somebody? But it happens. If you don't have that spirit, you don't belong in one of these cars.

Dad told me when I first started racing that I had to do exactly what I felt like I had to do to take care of a situation.

When I started in the sport, I'd been to dinner with these people; I knew how they felt about each other. I knew the guys my Dad didn't like that I did. It was different.

I still remember a race at Myrtle Beach. I was having trouble with Jack Smith. Jack had always been a guy my father could not get along with. I liked him, but he roughed me up pretty good at a race just before Myrtle Beach, and I told Dad, "You know, I don't exactly know how to handle this. I like Jack, but I am not going to let him run over me."

Dad said, "You're talking to the wrong person. I'll tell you right now, I don't like him."

Dad said if I had a problem with him, I couldn't start letting people take advantage of me no matter what age I was. I had to stand up for myself.

You meet all kinds of people when you're a famous race car driver.
Look at pro wrestler Andre the Giant's fist—it's as big as my head!

About 10 laps into the race, Jack rubbed all the way down the left side of my car. I said, "Well, here we go." I kind of caught him on the right-rear corner and sent him up the hill. After the race was over, I figured he'd be coming down looking for me. And he did. But he just looked at me and smiled. And we never had another problem.

That was just kind of the way you handled things. You didn't have to worry about what somebody felt about you, because before you left the race track, it was solved.

24

Walking the Competitive Edge

Buddy chuckles when you bring up the word "cheating." There's a simple formula for determining how many cars in a race any given driver is convinced is cheating. If the driver finished 23rd, then 22 cars were likely doing something illegal. His team, on the other hand, is merely taking advantage of anything not strictly and expressly forbidden. The rules tell you only what not to do. They don't say anything about what you can do.

There's an art to looking for the competitive edge, pushing it just as far as you think you can and getting by with it. I'm sure Bobby Allison and Cale Yarborough and Richard Petty will scoff and laugh at this, but I can't remember a time anybody ever told me that my team was doing anything that might have been in the gray area of the rules.

Well, no, I take that back.

When I was driving for Hoss Ellington, they did bring me something and show me they were going to do it. They had to hide it from Hoss, because he was so honest he would tell Junior Johnson, and Junior would tell NASCAR.

If there was anything in there, I didn't want to know it. It'd be hard to have a poker face if you knew it, but if you didn't know it couldn't hurt you.

Most of the cars I drove in my career were good enough that we didn't have to cheat. But we pushed it as far as we could toward that line.

Harry Hyde was not a cheater. They got on him one time at Talladega, they said the grille on the car was out too far. We were beating everybody by about 7 mph so they were hunting for anything.

Harry said, "Do you really think it makes a difference if you move that back a half-inch?" They said yes. He took the grille off and sent me out, and I went just about as fast I had before then. What they didn't know was he had a block on the radiator that blocked off the nose. The grille location had nothing to do with it. He had a little flap in there that would change the core in the radiator. Nobody said you couldn't do it. They came back and told us we could put the grille where we wanted it.

When David Pearson was winning all of his races, he was not cheating. When Cale won all of his he wasn't cheating—no more than anybody else. Cale was driving for Junior Johnson, and I will tell you that of all the people I've ever met in Winston Cup he would take it closer to the line because of his desire to win. He felt that if everybody else had an edge, he had an edge.

Junior was one of my heroes as a car owner. He showed up one time at Nashville with a Chevrolet Monte Carlo, and everybody said, "Well, Junior has stepped in it this time. That's the ugliest car I've ever seen." The windshield didn't look right, and the way it was built was such that we kind of laughed at him. We weren't laughing after the first practice, though, because he had us covered. They put a whipping on us, and pretty soon everybody was building one of those old wide Monte Carlos. You had to if you wanted to stay in the game.

Junior brought a car to the race track one time, and it was so bad they called it the "banana car." The nose was dragging the ground and the tail end was 10 feet in the air. You never saw anything that ugly in your life, but it was very fast. NASCAR crawled all over that thing.

Donnie Allison on the outside of row 1 and me on the pole in the Gray Ghost in the 1980 Daytona 500.

When I first started my race team with Danny Schiff, we took a car to Daytona one time. When we unloaded it, a NASCAR official said, "Buddy, I'll tell you what I am going to do. I am going to put the template on this car, and if you think I should let you run it, I will let you run it. I will leave it up to you."

He ran everybody else out of the template room and put the template on our car. It was 4 feet off in some places. It missed the template so bad. Robert Harrington was my crew chief, and I knew he hadn't done it on purpose. It was so bad I couldn't even argue about it. The other car we had fit the template like a glove, and we finished third in it. We must have got a template off something else, and that might have been what happened with that "banana car." But I don't think so.

Smokey Yunick was one the great mechanics in NASCAR history, and he and I always talked about why we didn't get together and go racing. He had worked with Fireball Roberts, who was my hero, and I loved Smokey to death. But the situation was just never right. When he wanted me, I was running for Ray Fox, and they were pretty big rivals. Their shops were about a block and a half away in Daytona.

At one time Smokey was unbeatable. His cars were the best. My dad drove one in Darlington one time. Dad was usually pretty colorful in how he described things, and apparently this car was a little stiff and hard to manage. We were going home from the track and I asked what the car was like to drive. He said it was the only car he'd ever been in that could spin the right-front wheel. It had so much power it would still be pulling at the end of the straightaway.

The great Smokey Yunick story is that one time in Daytona the NASCAR inspectors were on him about having an oversized fuel tank. They took the tank out and measured it, and it was fine and they gave it back to Smokey. He just laid it inside the car, didn't even hook it back up, and fired the car up and drove it back to his shop in Daytona Beach. They asked him about it later, and he said he could have driven it to Jacksonville if he had to. The rulebook didn't say anything about how much fuel line he could have or how big the lines could be. He probably had 120 feet of hose in there, and it was big old round hose.

About 90 percent of what Smokey had was a year or two ahead of everybody else.

NASCAR hired Gary Nelson to be the chief inspector, and everybody always said the best way to catch a cheater was to be one. He would know how to catch a cheater. When he was a crew chief, he could probably tell you some stories about going to the edge. His team won a summer race at Daytona with Greg Sacks in an "experimental" car, and that thing was like a bullet.

I would say they put Gary Nelson in the right job.

25

Minor Brain Surgery

Though he would drive in 17 races afterward, Buddy's days as a competitive driver effectively ended when he faced a life-threatening situation after a wreck in the 1988 season. After the wreck, though, Buddy pressed his luck until a scary moment on a road course in upstate New York finally got his attention.

I got hurt in Charlotte, of all places, in 1988. It all started for me there, and that's where it ended.

Eddie Bierschwale had not had a great day, and Bobby Allison and two or three others were battling like crazy up front. They went down into the corner in Turn 1, and something happened to Eddie's car. He turned around, and I slowed up. Bobby didn't have that option. He came across and hit me and turned me head-on into the wall.

My helmet hit the bar that goes to the right front with enough force to actually change the configuration of it a little bit.

That was the first time I'd ever been knocked out. I got the car stopped somehow; I got out of the car on my own, and I don't remember doing it. When I came to, they had me on the stretcher going toward the hospital. I kept complaining about my neck hurting. They checked me for a neck injury and said it was okay.

What I didn't know is I had started bleeding on the outside of my brain.

I was just trying to finish the year out. When you own a race team, you do some dumb things. I eventually got to the point where I walked to the race car and had to hang on to my son. I was just wobbling. But when I turned to the left I felt just fine.

The first time I knew I had a problem I was running Phil Parsons at Talladega on a restart right near the end. He slowed up a little bit, and I creamed him. I just ran right into him. I said, "That was pretty stupid, but I didn't see him."

When we went to Watkins Glen, the road course, that was the first time I had to turn right. By that time, though I didn't know it, I had a blood clot the size of your fist on the right side of my brain.

I went down the back straightaway and just blacked out. I woke up and I was just stopped. I looked around and said, "What in the heck am I doing down here?"

I told myself it was something I'd had for breakfast. I had been having vision problems and tremendous headaches at night, but I was still talking myself into continuing on.

After I fired the car back up there at Watkins Glen, I started back up the hill, and the road looked like it was just going all over the place. I thought, "Whoa!"

The first person beside my car when I got in was Dr. Jerry Punch, an announcer for ESPN who also is a doctor in the emergency room at a Charlotte hospital. I told him I was having all kinds of trouble and didn't know what was wrong. I told him I'd just passed out and was dizzy and was having those headaches. The top part of my vision looked like there were sparkles in front of me or something.

He said, "You've got an aneurysm or something." He told me who to go see and to go right then. I put Morgan Shepherd in the car, but I stayed through the race on Sunday. I drove the crew home that night and went over on Monday morning to see this doctor Jerry Punch had told me about, a doctor named Jerry Petty.

That was the other thing. When Jerry Punch told me to go see a doctor named Petty, I said, "Geez, the Pettys have been working on me all these years and now one's going to get inside my head?"

Anyhow, I checked in at 8 that morning with Dr. Petty, and I feel like that's the reason I am here. They did a CAT scan, and by 11:30 I was in the operating room. He told me they needed to do brain surgery, and I almost passed out.

I called Tom Higgins, the motorsports writer for *The Charlotte Observer* and one of my best friends, and said, "Higgins, you ain't going to believe this. I am going to have minor brain surgery."

He said, "Baker, I hate to tell you this, but there is no such thing as a minor brain surgery." He thought for a minute and said, "But in your case, there just might be."

It still didn't register. I told Dr. Petty I was going to go up to the shop and lay out the week's work for everybody. He said, "You aren't going anywhere. How you're standing up, I don't know."

I said, "I am a big boy; just lay it on the line."

He said, "Well, we think we can get you back."

I said, "What's the down side?"

He said, "Quite frankly, you could die." Just like that. That's when reality hit.

He had medical book sitting about 18 inches away from where he made the cut into my skull. He told me later that the pressure against my brain was so great that it moved the pages in that book when he tapped into my skull.

I thought racing was tough. That was the toughest thing I ever had to fight back from. The fight to get back to normal —not to racing, just back to living normally— was rough. I had moments where I had seizures and various other things that prepared me, probably, for things I am going to face later in life. There were times when I just thought, "I don't care. Nothing can be worth all of this."

I'd know what I wanted to say but couldn't say it. I remember the first time I went to the bathroom by myself. There was this woman who took me in there, and that got old in a hurry. So I decided to go on my own. I went stumbling in there, and when I went by the mirror they'd only shaved one side of my head. I saw the side that had hair and thought, "That's not so bad." When I came back by I could see the side they worked on. It scared me bad. I thought "Jason" had got into my room. I actually recoiled in horror at all of those horrible looking staples and discolored skin. Lord, I thought I wasn't going anywhere ever again.

They said they'd let me out of the hospital if I promised I would go up to my house on Lake Norman and relax. I was sitting around up here and decided that I needed to get a driver for our team. Darlington was the next week, so I took it on myself to go down there and talk to somebody. I went to Darlington and talked to Greg Sacks.

It was 96 degrees that day, and I started to get very, very cold, freezing cold. I knew something wasn't right, but after promising everybody I'd behave, I jumped in my car and all of a sudden I was absolutely burning up. I turned the air conditioner wide open and drove three-quarters of the way home before deciding I wasn't going to make it. I should have been on a leash. I finally did make it back, but that set me back another three months getting over the damage I'd done.

Then I started thinking about being out of racing at least a year. I wanted to run some more to prove to myself it was time to quit, but the bell had rung.

I am glad I ran more races, because I answered the questions for myself. But it was still hard to walk away.

The last three races I ran in 1992 were in cars owned by Derick Close. The last shot I had at doing something was in the Daytona 500 that year. We went down there and didn't even have a motor. We borrowed a motor from Bahari Racing and got an intake from Ken Schrader and a carburetor from Hendrick Motorsports.

We finished way up in our 125-mile qualifying race, and Benny Parsons came up to me and said, "Buddy, do you have any idea what you just did? That car was 58th fastest in qualifying, but you went out there in it and got yourself right up front for the Daytona 500. You could win this darn thing."

Schrader had given us the intake, but after I passed him late in the 125 he told the guys working with him, "If you ever give that guy anything else, I will fire you."

When the race started, that car was running pretty good and I was thinking, "Hey, this could be something here." It was flying. If the motor had stayed in it we could have done pretty well, but the motor blew.

When it was finally over for me, the hard part was the realization that I couldn't do what I'd been doing my whole life and that I may never be able to do it again. It was the most empty feeling I've ever had. It was like everything I'd ever stood for was gone, like a part of me had been cut off.

You begin asking yourself what you're going to do the rest of your life. The answer was I was going to own a race car and hire drivers, and I started wondering if I could do that. The only driver I'd ever had was a fellow I knew I couldn't fire —myself.

When I was there every day and working with it, my race team was pretty consistent. When I got out of the car and wasn't at the race shop every day to take care of business, it wasn't the same.

Dale Jarrett wanted to drive for me, way before he had a chance to get in the kind of cars he eventually got into. I sat down with Dale and Ned, his father, and told them, "Right now, that would be the biggest mistake you could make. You're capable of being a winner in this sport, but my race team isn't capable right now. By the time you got good enough to win, I'm afraid the team might not even be good enough to continue. So I won't hire you."

But the first time I watched my race car go around the race track without me in it? That was kind of like the first time somebody comes to your door and says, "We'd like to sell you a burial plot."

26

A Eureka Moment

When Buck Baker retired from NASCAR competition in 1976, he didn't exactly park his interest in racing. It was a dream of his to start a school to teach potential racers their craft.

While levels of participation vary, the Buck Baker Driving School is not one of those deals where someone runs six laps as a passenger in a car somebody else is driving. Winston Cup stars such as Steve Park, Jeff Burton and Ward Burton have trained at the school, which is based at the North Carolina Speedway in Rockingham, N.C.

Another young driver who came to Buck's school was a fellow from Indiana named Jeff Gordon. By the time he came to the Sandhills region of North Carolina, Gordon was already a star in the U.S. Auto Club midgets and sprint cars in Indiana. But it was at Rockingham where his path to the stardom he enjoys today began.

We knew about Jeff Gordon from ESPN's *Thursday Night Thunder* show, which showed sprints and midget cars racing from all over the Midwest.

We'd watch Jeff and Rich Vogler go at it week after week. Jeff had this little mustache but still looked like he was 12 years old, and he was going out there in a sprint car and running with Vogler and others who were among the best in the world, and he was beating them.

I've heard people say Jeff never had to earn his way. They're crazy. He started racing when he was just a kid, and in go-karts and sprint cars, he was the best. When he started in the Busch series, he'd already won 400 or 500 races. He paid his dues and worked his way up the hard way.

My dad ran a driving school, and Jeff came down there to North Carolina Speedway in Rockingham.

One night on *Thursday Night Thunder* he said he was going down there. I called Dad and said, "Hey, whatever you do, this kid is the real deal. Do whatever it takes. Put the red carpet out." I'd become a fan watching this kid driving on TV. He was supposed to be the next A. J. Foyt, the guy who goes to Indianapolis and does it all.

Jeff Gordon didn't stay in the regular class long at Dad's school. We always kept a couple of cars down there that were really race cars, lots of horsepower and set up to handle well. He was in the regular class for about two hours, and then we moved him over.

He was so good there was a guy in the same class at the driving school, Hugh Connerty, who was with the Outback Steakhouse chain and who was going to buy one of our cars and enter it in the Busch race at

Dad and I celebrate my victory in the 1973 World 600 at Charlotte.
Do you think they made the trophies big enough for us back in our day?

Rockingham. This guy was so taken with Jeff's ability he decided to let Jeff drive it.

The night of his first day at the school, Jeff called home and talked to his mom. The door was ajar and somebody heard him.

He said, "Mom, I found what I want to do the rest of my life."

We've had a lot of people—the Burton brothers, Steve Park. It was my father's dream, and people like that made it legitimate. It wasn't a ride-along deal. You went out and you learned to drive. When Jeff got there he was already a great driver, and he liked these cars. That was the direction he wanted to go.

Jeff went on into the Busch series and then signed with Rick Hendrick to drive in Winston Cup. He raced one race in 1992, the last race of the season. I was still racing then and doing some testing programs too. The car that Jeff Gordon won his 125-mile qualifying race in as a rookie at Daytona in 1993 was a car I'd worked on at Talladega with Ray Evernham. It was one of Ken Schrader's cars. They didn't have time to finish a new car for Gordon, and Ray and I worked on it before they took it to Daytona and won with it.

27

Aren't You That Guy on TV?

Buddy didn't step out of his race car and straight into the television booth. In fact, there were several years when he still had one foot in each—he was still racing and owning a race team while working on some race telecasts.

Buddy would eventually go on to join the CBS network telecasts of the Daytona 500 and other major races, becoming a favorite of race fans nationwide. In 2001, however, when Fox and NBC took over rights to Winston Cup and Grand National events, Baker was not included in the new announcing teams lined up for those broadcasts. He is still the analyst for TNN's coverage of American Speed Association events and also makes a number of speaking engagements each year for a large tour company that brings in fans for NASCAR events.

I ran one of the few in-car broadcasts that have ever been done. I did it in the Winston Open at Charlotte for TNN. I started last and worked my way up to the third spot.

I told them, "Boys, I am enjoying the broadcast, but I am going to tell you this is getting serious now. I am going to get off here and go to work." But the motor blew up a couple of laps later.

I started doing racing on television back when nobody wanted to do racing.

We went to Riverside, Calif., in 1987, and I got Irv Hoerr to drive my car.

Irv wasn't used to the Winston Cup car. He was sliding the wheels going into the corner, and I didn't realize we were back from commercial break. I was on the radio telling my team, "Will you tell Irv it's not against the rules to pass somebody here?"

My first race for TNN was at Dover with Mike Joy and Phil Parsons. I did that first race, and I told Patti Wheeler, who is Humpy Wheeler's daughter and who was the producer for the telecast, that I didn't like it. I felt like I was talking about what was obvious, things that people could see. But Patti asked me to do one more because they liked what I did.

I told Mike Joy it was boring to just sit there watching people do what I'd always done. He told me it might help if I went into the garage area and got to know the people and why things were or weren't working. He was telling me I needed to tell the people something they couldn't just look straight at.

That helped make it a job for me, something to work on. Before very long, I loved it.

When CBS called me to do the Daytona 500, that was just amazing. I think the first one I did was 1995 or '96. When Eric Mann called me, it was like winning the race down there. It was just too nice. That group from CBS stayed in the best places and did the best things for me. That was just the best.

I just tried to be honest about what I saw, and sometimes that rubbed people the wrong way.

I know my answer probably griped some people when they asked me about it, but I said, "If you can't tell the truth on TV, you shouldn't be there. If you're scared of losing that job to the point you're going to lie to the American public that's sitting there looking right at something, why do it?"

I was very sympathetic with the drivers, and as a former driver it figures I would be. But if a guy just turns somebody around, that's what he did. You can dress it up and change it around but it's still the truth.

I loved doing television for 12 years and got to work with a lot of talented people. Here I'm getting ready for a TNN telecast with Eli Gold (center) and Dick Berggren.

One guy got on me one time for saying he wasn't trying very hard. I said, "You were four laps down; were you giving it 100 percent?"

He said, "Not right then; the car wasn't right."

I told him, "That's all I said."

One time at New Hampshire, Dale Earnhardt turned Sterling Marlin coming off Turn 2, and Sterling waited on him and ran into him. I think the comment was, "He who lives by the sword dies by the sword."

When I am doing a race, whether it's Winston Cup or the ASA, I have a job to do. A stiff arm is a stiff arm. If there's a ruling on the race track that I think is bad, I will say it's bad.

NASCAR has said things to me, sure, but I've never been one of those people who goes to bed wondering, "Did I mess up?" I've never had to back up on the truth.

I have a passion for life, and in anything I do I am going to use that. When I finally decided the better part of my career was behind me and I could no longer represent what I stood for over all those years, I tried to take that same attitude into the broadcast booth.

Every day at my house, I have to carry my mail with two hands because I get letters and notes from people telling me they like what I did. That's such a bonus. I had never really been anything in my life except a race car fan, and to be able to turn that around and have people enjoy what I did and know I did it pretty well, working for cable and the networks, that was wonderful.

People ask me if I am bitter about what happened. Not a bit. I had 12 great years doing it. I made a lot of good friends and learned a lot about the sport. I am totally a race fan because I know how hard it is to make it.

In 2000 after the last Daytona 500 we did on CBS, a guy named Steve Rushin wrote an article in *Sports Illustrated* saying my commentary was "delightfully indecipherable."

It's always good to have something written about you in *Sports Illustrated*, I guess. At the time, I was doing a column for their online site every week. He meant it one way, and for me it was taken another. I didn't appreciate it and still don't. But should I alter my life because of one person's opinion?

Did I know what I was doing? CBS thought so, apparently. How many more people know me than know who that guy was? If Steve Rushin walked up to me on the street, I wouldn't know him from Adam's house cat or from anybody else who ever loved racing.

He has his opinion and he's got a right to it. I didn't change the way I did things because of anything anybody ever said, and I don't guess he did either.

He wrote about me having a Southern drawl. Everybody sounds different. On Lake Norman where I live, I sound pretty normal. One man in New York thinks having a Southern drawl is a little bit of a handicap, but if he came down here I guess we'd laugh at him, too.

28

A Lost Generation

    The 1992 Winston Cup season was a year of transitions in the sport. Richard Petty drove his last race and Jeff Gordon drove his first race. It was also the final season in which Buddy made an official Winston Cup start—he was in three races, bringing his career total to 699 races.

    Alan Kulwicki edged Bill Elliott and Davey Allison in a dramatic championship race in '92 that came down to, literally, one lap that Kulwicki managed to lead and Elliott didn't. When it was over, Kulwicki was perhaps the most unlikely champion in stock-car racing history while Elliott's car owner, the legendary Junior Johnson, started toward his exit from the sport.

    Allison, driving the No. 28 Fords that were directly descended from Harry Ranier's team that gave Buddy so much of his success, seemed destined for greatness in the sport that would soon take a tragic toll on him and his family. He and

Kulwicki would be killed in separate aircraft crashes the following year.

It's still very painful for me to see Bobby Allison. The things that family has been through.

I raced against all of the Allisons. Every one of them was a winning race car driver, and they would have won in any era. But Bobby, he was the world's worst about running up on somebody who was a lap down and bitching and bellyaching about them not getting out of the way. But he could be 20 laps down and he was the hardest man on this earth to pass, absolutely impossible.

Nobody has ever loved racing more than Bobby Allison did. No matter how much you said about him, you could never say too much about how great he was. He could take a bad car and run way up there. He won in an American Motors car, a Matador. How tough was he? Who else could have won in that car?

Had Davey Allison made it and Tim Richmond and Alan Kulwicki lived, the complexion of our sport would be totally different than it is. As great as things have been, they would have been that much greater.

We lost four or five drivers right there at one time who would have been absolutely spectacular. I know Davey would have won a couple of championships by now. Somebody would have signed on with Alan Kulwicki and he would have been on with it.

I raced against all of those guys, and it would have been so much richer with all of those guys in the field.

The battles Richmond and Dale Earnhardt would have put on would have been incredible. Davey Allison would have been in there, too. Alan Kulwicki probably would have sat there and watched them and outfoxed them all.

I think Tim Richmond, and this is no reflection on anyone else, when it came to driving a race car that was out of shape, was the best one I ever watched. He could recover from almost anything. No matter how hard he went into a corner, he could hold it. I watched him go into a corner at Riverside one time, and he used every rock and every inch of that race track. Every time you'd say, "There he goes...no, he brought it back." As far as natural ability, he was just about the best.

We'd watched Richmond at Indianapolis and in the midgets, and we knew he was good. If there was one downfall for him at all, it was that he almost had too much talent. I mean, it just wasn't like it was work for him. He got himself into situations a lot of us couldn't get into because his father was wealthy. Tim was kind of a playboy and he enjoyed everything.

He had one of his driver's suits done up with wings under the arms like something Elvis would have worn. He played to the crowd and loved that kind of stuff. He'd spend $2,000 on a tuxedo and live on yachts. He wanted to be an actor.

I didn't have a whole lot of close dealings with Alan Kulwicki. I don't think very many people ever did. He was so private. His father used to build motors for Dodge that the factory would send to us. The first time Alan came to Winston Cup, none of knew him. He qualified up front and crashed out, but we found out pretty quick that he was dad-gum smart.

He just wasn't the kind of person who'd call you up and say, "Let's go eat lunch." He lived one thing, and that was racing.

We went to a Christmas party one time, and he was there. He stayed about 15 minutes. He said, "Well, this was nice but I'd better go. I've got some work to do."

It was an incredible feat for him to win that championship in 1992. Incredible.

Davey Allison made my life pretty hard. I knew him from when he was just a kid—I have run him away from my car many times. He would come down there heckling me about what his dad was going to do to me, the same way I had done when I was a kid to all of those other guys.

When I got to know him as a man, we traveled together doing what they called Legend races. Harry Gant and Darrell Waltrip and a bunch of us would go to a local track and drive the cars there in a feature race.

Davey found something out about me one time. He came up behind me just before qualifying at Atlanta. Before I went to qualify, I was like somebody who had 220 volts of electricity flowing through them. I would get really up. I had been in racing for 30 years without anybody finding out how ticklish I was. He came up behind me and grabbed me in the rib cage, and I kicked the guy in front of me about five feet in the air.

"Oh," Davey said, "you shouldn't have let me know that."

From then on, he would sneak up behind me and gouge me. He and Adam Petty were the same way. I met Adam when he was a kid and got to know him as he got up into the Busch series. They were both just rays of sunlight to be around.

Adam and I were in a fishing tournament one time, and we were coming back from Palm Beach, Fla. I got on the airplane, and I was sitting in first class and he was back in coach with some of his buddies.

I told the stewardess, "You see that tall, skinny boy back there? That's Richard Petty's grandson and Kyle Petty's son."

She said, "What's he doing back there?"

I said, "Can we get him up here?" and she said yes.

She went back and got him, and like all young guys he'd been out of money for about three days. He came up there and sat down in that big seat and looked back at his friends and made a face at them.

The girl came back through there and said, "Can I get you something to drink?"

He leaned over to me and whispered, "How much?"

I told him it was free and he wanted two Cokes. Everything they brought him he'd hold it up and show it to the guys in the back to mess with them.

When I heard about Davey's helicopter crash, I first heard they were taking him to the hospital, l and that gave me a little hope. After that, it was like a bad dream, like when Fireball Roberts got hurt. I wanted to go see him, but I didn't want to see him like that.

I did go to Davey's funeral, and something really unusual happened. It was the most eerie thing I have ever seen in my life. Midway through the service, a beam of light came in through the top of the church, and—you can ask anybody who was down there—came through a skylight and directly on Davey's casket. You could hear people gasp. I have never seen anything like that in my life.

That was one of the last funerals I attended until my father died.

I am not sure I could have coped with going to Adam's funeral and seeing Richard and Kyle after having lost my mother a couple of years earlier. I know you should go out of respect. I raced for them, but I just couldn't handle it.

I was doing the race at New Hampshire the weekend Adam got killed. For months after that, every time I would go by Kyle at the race track I'd want to go over and say something, or just hug his neck. But I never knew quite what to say or do.

29

A Trained Observer

Buddy spent more than 30 years driving race cars and then was, for more than a decade, a television analyst on Winston Cup and Busch Grand National events. If you've come this far in this book, you know Buddy isn't afraid to tell you how he feels about anything. That's especially true when it comes to a sport he still dearly loves.

One thing I get tired of is reading in the newspapers on Monday that somebody was a sixteenth of an inch too low after a race. During the race they let them work on the cars and raise and lower them with the jackscrew. If your car is way off in how it's handling, the way you get a better balance is to lower one corner of the car. A sixteenth of an inch? An eighth of an inch? Come on.

I could see it if it was a half-inch. You've already been through a prerace inspection. Why do you go through that?

I can understand checking the spoiler heights. But if they're not within some certain tolerance, don't give them any money and don't give them any points. They're out. And you'll stop it.

You have to have common sense. Don't let a guy win the race after passing every kind of inspection there is, then say he wasn't legal after the race but you're going to let him keep the win.

Some of the things NASCAR has let happen to the race cars, I don't agree with. People love this sport for the enjoyment of watching people compete. When the cars get to a point they're so streamlined the drivers can no longer compete and make the fan feel special about being there for that day, I think you've hurt the sport.

The idea of a common template sounds like a perfect solution, where the fans don't have to worry that one car has an advantage over another and the race is won by the best driver and the best team—how a race should be decided. It also would seem that common templates—meaning that NASCAR would measure all makes of cars exactly the same way—would eliminate the ridiculous expense of going to the wind tunnel and changing the cars six or seven times during a season as NASCAR rules change.

But it's not that easy. Cars would still have front sections, the noses of the cars, that are different. The Dodge would be different from the Pontiac, which would be different from Ford and Chevrolet. The noses of the cars have to look different so the car manufactures can retain some sense of identity for their product. So race teams are going to work on those noses to make their cars better. If one car gets better, the others start complaining and the beat goes on. You're right back where you started.

A lot of things just don't make sense. Why do they put restrictor plates on cars at Daytona and Talladega and then let the teams work on their intakes? What are you doing? Why let teams spend all winter working on getting more horsepower out of restrictor plate motors when the purpose of the plate is to limit horsepower? Give them an intake and say, "This is it and if there's a mark on it, you're out."

As a car owner I spent a lot of money to get more horsepower when the whole point was to limit it.

I don't believe that anybody has ever gotten a "company plate" at a restrictor plate race—an oversized plate given to him so he's got an advantage. People have suspected that's happened, but I don't believe it has. To remove all doubt, though, if we can put a man on the moon, we certainly can make a machine that issues the plates randomly. When a guy walks up, the machine spins around and whatever plate comes out, that's the one a team gets. Why does NASCAR have to hand you one? If they had that machine wouldn't that take care of it?

Don't get me wrong, I am absolutely in love with driving race cars at Talladega and Daytona without that plate. I loved being able to get a run on somebody and make that pass and if I'm coming right there's nothing to stop me. You still had a chance if you were running third to pull out and get a run on somebody and win the race.

I love that, but I was just in front of Bobby Allison when his car flew off the race track at Talladega in 1987. I had passed him coming off Turn 4 and when I went down through the dogleg I looked in the mirror. Bobby was pretty hard to pass, and he would try you back on the inside. I looked to see where he was and when I did I saw the bottom of his car going the wrong way, up toward the flagstand.

I was fearful for Bobby, but at the same time I was thinking that could be the end of our sport. It looked to me the car was going over the fence, and if it had it would have taken a section out. It would have gone through the people all the way down to the first corner.

I am for anything that lets the fan leave the race track healthy. You don't buy a ticket to put yourself or your family in danger. I don't know if the restrictor plate is the answer; it may not be, but it's way ahead of anything else anybody else has come up with.

I am 61 years old, and if I went to Talladega right now in Rusty Wallace's car and put the whiz-bang motor in it, I don't think there would be any problem in me running it at 240 mph. I believe that. Bill Elliott qualified at 212.809 mph at Talladega in 1987 with about 620 horsepower. Unrestricted motors have about 840 horsepower now. The tires are twice as good now as they were then. The cars are down lower. I think 240 mph might be a conservative number.

We've been down there for test programs in the winter time when it's like 17 degrees in the morning, and we slicked the car up one time and the car blew up coming off Turn 4. I coasted past the start-finish line. I coasted maybe 200 yards and still ran 219 mph. God knows what it would have been if the motor hadn't blown up. That was a long time ago. You could do 230 mph now and it wouldn't even be saying anything.

I don't like to say bad things about other forms of racing because I am a fan of all race, but the wreck the Indy Racing League had in Atlanta in 2001 where one car just exploded on the backstretch? Had that happened with a Winston Cup crowd in those grandstands, I don't know how many people we'd have lost there.

It's not about how fast. It's about the competition you have. From the flagstand, you can't tell 200 from 210. You can't look at a car and tell that. Watch the International Race of Champions. They run 10 mph slower than Winston Cup cars, but they don't look it. It's competition.

I was the first driver to run 200 mph at Talladega in 1970, and I went back down there in 1975 when Mark Donohue ran 221 mph in a Porsche to set the closed-course record. It just didn't look that fast on a track that size.

For the most part, when I drove the drivers respected each other. The stuff these guys do today, cutting each other off and stuff like that, I will tell you right now I would have turned them around. There's no doubt about that or arguments. I would just tell them that if I'm here and you come across, I am still going to stay in my position.

If people did that to Cale Yarborough, they might as well go ahead and run into the wall before Cale got back to them. Richard Petty, there's another one. He's a nice guy, but if you didn't respect him he wouldn't take it. There wasn't any blocking or any of that stuff. If you cut Cale off, you were wrecking. If you did that to Bobby Allison, Bobby was a mean son of a gun on the race track.

If you wrecked a guy to win a race, you knew you were going to get it back. And we didn't have the problems they have now. Had that kind of stuff happened in our day, cutting people off and getting by with it? That stuff amazes me.

NASCAR, however, has done a remarkable job with making the sport what it is.

If I had to sit down right now, as much as I am still involved in racing, and tell you the top 10 drivers in the CART series, I could not do it. If somebody said a driver's name, I might know what circuit he runs in, maybe. But there was a time you could not get a seat at a CART race.

The pushing and the pulling and the greed all hurt that kind of racing. People have talked about changing the format of Winston Cup, dividing it up into two divisions like and East and West—you want to watch it die, do that.

The fans pay each week to see the best drivers compete. That's what has made the sport what it is. The fan knows he can go to the Cup race and see every driver he knows and know those drivers are going to race that day. Take them and divide them up and then say you're going to have a Super Bowl race at the end of the year? That will knock it out right quick.

There are a lot of great, great drivers in racing today.

I've been fortunate to have the chance to work with Ryan Newman over the past few years, and he has been terrific. I told him one time that if he wanted to have somebody to tell all of the things he shouldn't do, then I was the man for that job.

Tony Stewart is one of the greatest drivers to ever come through here. He's just going to add to his story throughout his life if he stays healthy. I don't know how to explain this to people, but some guys are just winners and Tony's one of them.

The first time I ever met him, he had that look in his eye. Harry Ranier, the guy I used to drive for, brought him to NASCAR in the Busch Series. He brought Tony to me and wanted me to talk to him. I didn't know him, but Harry said, "Buddy, he is the real deal."

I covered the Copper Classic in Arizona one year and Tony won everything out there. Well, he won the midget and sprint races going away and then climbed into a super modified and started dead last and lacked two inches of winning that.

Jeff Gordon was the same way. When I watched him on Thursday nights racing those sprint cars on ESPN, you could just see it. I don't even like to watch television much at night, but I'd hurry home from the track on Thursdays to get there in time to see him drive in the heat races. Jeff looked like he could shave with a cat, but he was kicking butt and taking names in a sprint car with 850 horsepower as a teenager.

I love Dale Earnhardt Jr. I think he could be as good as anybody who's ever driven a car, but I don't know whether he gets the proper amount of rest and keeps himself in the kind of shape he needs to be as good as he can be, at least not yet. The only question about him is does he really want it? At times, I think having a good race car isn't as important to him as maybe it will be later on in his career. We see signs of his potential brilliance now, but it's the tip of the iceberg. When he matures a little more and 100 percent puts his nose to the grindstone, look out.

The first time I ever really noticed him was at the exhibition race in Japan when he took a show car over there and beat some of the best drivers in the world. Richard Childress didn't believe it. He knew where the car had come from.

Some of these guys out there have had a little success, but I don't see the quality you need to be one of the great stars like a Pearson or a Petty, somebody who sticks out. Some of these guys haven't learned yet they can't win by being stupid. They can't win with the front end of the car folded up. You don't see cars come to the winner's circle with the radiator back on the race track.

Some of these guys have a glimmer of talent and win a race and then they try to show everybody how not to win. You can't tell them anything because they won a race. Show me about 15 wins and you've showed me you've learned the formula.

One thing that's bad right now is that you can do nothing and still make a lot of money. You can be a bad boy and do everything wrong and still make $1 million. But racing is like playing King of the Mountain; you have to fight to get there and then you have to fight like hell to stay there. If you want to stay on top of the hill, you have to really fight for it with everything you have.

I hear people complaining that the drivers won't sign autographs for anybody who wants one. Richard Petty did that, sure, but he couldn't do it now. He'd be 114 years old before he left the track at Charlotte.

Still, the drivers are more isolated than they need to be. When I was a broadcaster I could walk up into the trucks because I was a competitor and talk to those guys.

That's a big advantage over somebody who has to go to the back of the truck and ask the public relations person go up there and see if the driver will come out.

I've watched a driver do that to Ned Jarrett one time and it infuriated me. The driver told Ned he'd be out in 20 minutes and Ned sat on the back of his truck. Twenty minutes later the driver walked back there and told Ned, "No." Just like that.

That frosted me. I told the driver about it. "Where do you get off?" I said. "Without that guy, you would not be doing what you're doing today. Him and people like him. And, when you get down to it, more people know him than will ever know you."

The fans need to understand some things, too. When a driver is out in public and somebody comes up and asks him for an autograph, I think he has an obligation to accommodate the fans as much as he can. But in that garage area, it's different.

The only way you can be as good as you need to be is to take the time to talk to your crew and translate what the car is doing on the race track. When somebody stops you and starts talking, a fan or a reporter, I don't care how smart you are, you lose part of the information you need to translate to the crew chief.

There are far, far too many people in the garage area at the time when people are doing a lot of important work. If they have a tour through there on race morning, that's one thing. But when those teams are working and cars are backing out and people are running this way and that to get springs and shocks, that's their office in there.

One thing I think the drivers ought to do more is just go out in public more than they do and be around folks. Go to the shopping center or the movies or the grocery store and people will get used to seeing you. It won't be such a big deal because they will see you all of the time.

Sooner or later, NASCAR is going to have to restrict the time people can be in the garage area. There's going to have to be a time for the drivers to talk to the media and to the fans, scheduled the same way they schedule time for practice. You can't isolate the drivers from the media and the fans, but you can't let so many people in the garage area that these guys can't do their work.

The money's out of control, too. If the rules stayed the same long enough, maybe you could keep a car a couple of years and just make minor modifications and keep using it. But if you're going to have a whole new set of templates next year, how do you set a budget for that? If they say they're going to change the engine blocks and run smaller engines instead of using restrictor plates, how much money does that cost everybody?

Back even when I was racing, we'd go test for two dad-blame weeks at Daytona and they'd change the rules the week before we went back down there to race. All of that testing wasn't worth a flip. You go spend money like crazy and they decide they didn't like the way the speeds lined up.

It should be more consistent as far as the rules. Print the rulebook and that's it for the year. Do away with all this complaining about "You did this for the Ford," or "You cut this from the Dodge."

If somebody made me king of NASCAR for one day, I don't know what I'd do. NASCAR's job is pretty much impossible now. There used to be one guy who made the decisions and that was it. Somebody would say, "This car is better and we can't beat it."

Bill France would say, "Well, you'd better build one then."

30

Too Fast to Live?

Buddy has been around racing long enough to see his share of tragedies. As has been discussed earlier in this book, his hero Fireball Roberts died from burns suffered in a crash at Charlotte, and Buddy's good friend Tiny Lund was killed in a crash at Talladega. Buddy was doing television on the weekends that Adam Petty and Kenny Irwin were killed in crashes at New Hampshire International Speedway.

Some consider racing too dangerous and wonder why drivers would risk their very lives to compete. For Buddy, those kinds of questions aren't hard to answer.

There's nothing dangerous about living your life.

If I could push a button and become 20 years old again, I would do exactly what I've done all my life. I would be a part of racing.

There's no part of my life that I can remember when I was more alive than the minute that green flag came out and I started a race.

From left: Bobby Allison, a Union 76 racing queen,
Buddy Baker, Richard Petty, and Donnie Allison.

I've had things that startled me on the race track, but I've never been afraid. Even after I got injured and it could have gone either way—I could have died just as easily as I made it—the next time I got in the race car I had absolutely no thoughts of the wreck itself.

If you choose to be a test pilot, you compete against the equipment and against the unexpected. In my case, I loved competition to the point it became a way of life for me. I could not have had the same life I have doing anything else and been happy doing it.

There are times when you leave the track thinking, "What in the hell am I doing? What am I thinking? Why do I put myself through the aggravation of this thing blowing up with five laps to go?" But before you get home, you're thinking, "Boy, was that ever fun until it did blow up!"

You wonder why a tire couldn't make it three more laps. It was because it didn't, that's why. It's part of what you do. Things aren't going to go perfectly.

Even after I quit the sport, every year when the cars would be getting ready to go back to Daytona, there was a little voice in my head that went off and said, "I want to go!" It was the same voice that, when I was driving made me better, because every year when I was getting ready to test, it would say, "I hope I can be competitive this year."

Even after I'd won races. I'd go back to a track where I'd won the year before and think, "Gosh, I hope it runs that way again."

The sickest I've ever been in my life is right after a win. I get nauseous. You put everything inside of you into winning and then you do and it's all over. I would get sick on my stomach, every time. It would last for a couple of minutes.

It's like people who run a marathon. Look at them right after they stop. When I won the Daytona 500, I had to sit in the car for a minute because I didn't want to throw up in victory lane.

I can tell you that I love the sport enough to say I wouldn't change my life one bit. I could have done a lot of things in my life. But nothing ever suited me more than running out front. When you've got 42 cars stacked up behind you, when you're the best and everybody is chasing you, when you can hear the crowd roar over the sound of that engine—you can't buy that anywhere.

Even second place is not losing. It's about effort. As long as I gave 100 percent, that was as good as I could do.

I've always had the theory that if you stay around the top five enough, things will happen on the positive side. But if you're not in the top five, it's not going to happen no matter what you do. When you put yourself in that position and get passed in the last corner, you ask yourself what you didn't do right, but you don't feel like a loser. There are 50 people in the world who do this job at any given time in the sport, so you're a winner when you make that top echelon.

Now I would rather eat a green worm than lose. The competitive me is the guy who just hates that. The guy you meet on the street is a different guy. That driver fellow is a little spooky. I've had some of my best friends come up to me and speak to me before a race at Char-

Spraying champagne in victory lane after my long-awaited Daytona 500 victory.

lotte, and I've walked right by them like I didn't see them. It'd be like me walking into somebody's office on a deadline.

It's not about being stuck up; it's being stuck down. I'm ready. I'm about to go into a field of battle as far as I am concerned.

When they drop the green flag, my friends are working on my race car when I stop. I like all of the other drivers and I respect them, but I would do anything short of creating a wreck to win.

I have to be focused on how to change my line or tell them what they need to do to the car. I don't have time to worry about whether I was a nice guy before the green flag.

Most of the time I wouldn't even go out to eat on Saturday nights. I didn't want to hear anybody tell me that somebody else was running good—I didn't want to hear that. I was already getting into race mode. I'd have my food brought in so I didn't have to go out and be aggravated.

It's not unusual for men to like a challenge. The scariest thing in life, to me, would be to die at 90 years old and never have done anything you wanted to do. That's a waste.

I am happy; I always was. And that's hard for people to understand, why a guy would want to go 200 mph and take a chance on getting hurt. It's apparent that it doesn't always have to happen that way. David Pearson never spent a night in a hospital until he was out of racing.

You've been racing since you can remember. It's just what you do. If you own a grocery store and the door falls on you, would you open the store the next day? Absolutely. You might jerk on the door a little first, but you'd go on. That's your life.

I always understood that. I chose this and I loved it. If something happened on the race track, it happened.